Y0-BRD-751

NON-NEGATIVE MATRICES

NON-NEGATIVE MATRICES

An Introduction to Theory and Applications

E. Seneta

Department of Statistics, Australian National University
Canberra

A HALSTED PRESS BOOK

PHILLIPS MEMORIAL
LIBRARY
PROVIDENCE COLLEGE

JOHN WILEY & SONS
New York

QA
188
S46

First published in 1973

This book is copyright under the Berne Convention. All rights are reserved. Apart from any fair dealing for the purpose of private study, research, criticism or review, as permitted under the Copyright Act, 1956, no part of this publication may be reproduced, stored in a retrieval system, or transmitted, in any form or by any means, electronic, electrical, chemical, mechanical, optical, photocopying, recording or otherwise, without the prior permission of the copyright owner. Enquiries should be addressed to the publishers.

© George Allen & Unwin Ltd 1973

Published in the USA
by Halsted Press, a Division
of John Wiley & Sons, Inc.
New York

ISBN 0 470 77605 6
Library of Congress Catalog Card Number: 73 10891

Printed in Great Britain

To my parents

Preface

Since its inception by Perron and Frobenius, the theory of non-negative matrices has developed enormously and is now being used and extended in applied fields of study as diverse as probability theory, numerical analysis, demography, mathematical economics, and dynamic programming, while its development is still proceeding rapidly as a branch of pure mathematics in its own right. While there are books which cover this or that aspect of the theory, it is nevertheless not uncommon for workers in one or other branch of its development to be unaware of what is known in other branches, even though there is often formal overlap. One of the purposes of this book, through its aiming at breadth rather than depth, is to relate various aspects of the theory, insofar as this is possible.

The author hopes that the book will be useful to mathematicians; but in particular to the workers in applied fields, so the mathematics has been kept as simple as could be managed. The mathematical requisites for reading it are: some knowledge of real-variable theory, and matrix theory; and a little knowledge of complex-variable; the emphasis is on real-variable methods. (There is only one part of the book, the second part of §5.5, which is of rather specialist interest, and requires deeper knowledge.) Appendices provide brief expositions of those areas of mathematics needed which may be less generally known to the average reader.

The first four chapters are concerned with finite non-negative matrices, while the following two develop, to a small extent, an analogous theory for infinite matrices. It has been recognized that, generally, a research worker will be interested in one particular chapter rather more deeply than in others; consequently there is a substantial amount of independence between them. Chapter 1 should be read by every reader, since it provides the foundation for the whole book; thereafter any of Chapters 2–5 may be read independently of each other. For the reader interested in the infinite matrix case, Chapter 5 should be read before Chapter 6. The exercises are intimately connected with the text, and often provide further development of the theory or deeper insight into it, so that the reader is strongly advised to (at least) look over the exercises relevant to his interests, even if not actually wishing to do them. Roughly speaking, apart from Chapter 1, Chapter 2 should be of interest to students of mathematical economics, numerical analysis, and combinatorics; Chapter 3 to mathematical economists and demographers; and Chapter 4 to probabilists. (This last is believed to contain one of the first expositions in text-book form of the theory of finite inhomogeneous Markov chains.) Chapters 5 and 6 would at present appear to be of interest primarily to probabilists, although they contain almost no probability emphasis. In the key, and therefore highly unified, Chapters 1 and 5 all bibliographic notes and discussion, as well as exercises, appear at the conclusion of the chapter; this is also true of the short Chapter 6. In the less well-unified Chapters 2, 3 and 4, these appear at the conclusion of the sections themselves, for convenience of the readers interested in specific sections.

The author has attempted to provide generally simpler proofs for many of the well-known propositions, in keeping with the motivation for the book.

Various portions of the contents of the book have constituted parts of lecture courses to final year undergraduate and first year postgraduate statistics students at the Australian National University.

The author wishes to acknowledge the help he has received with materials from T. A. Sarymsakov, G. N. de Oliveira and J. Pitchford in particular; and wishes to thank the head of the Department of Statistics at the Australian National University, C. R. Heathcote, for providing moral, and arranging material, support. He wishes to thank also the ladies of the Economics Faculty, Mesdames J. Radley, I. Kewley, L. Ward and B. Palmer, for typing the manuscript with fortitude and perseverance; and his family for their patience.

Finally, thanks are due to my mentors in non-negative matrix theory: J. N. Darroch, who taught me about the finite case; and D. Vere-Jones who introduced me to the elegant extension of the Perron–Frobenius theory to countable matrices.

Canberra and Princeton, 1972. E. SENETA

Contents

PART I

FINITE NON-NEGATIVE MATRICES

1 Fundamental concepts and results in the theory of non-negative matrices

We shall deal in this chapter with square non-negative matrices $T = \{t_{ij}\}$, $i, j = 1, \ldots, n$; i.e. $t_{ij} \geqslant 0$ for all i, j, in which case we write $T \geqslant 0$. If, in fact, $t_{ij} > 0$ for all i, j we shall put $T > 0$.

This definition and notation extends in an obvious way to row vectors (x') and column vectors (y), and also to expressions such as, e.g.

$$T \geqslant B \Leftrightarrow T - B \geqslant 0$$

where T, B and 0 are square non-negative matrices of compatible dimensions.

Finally, we shall use the notation $x' = \{x_i\}$, $y = \{y_i\}$ for both row vectors x' or column vectors y; and $T^k = \{t_{ij}^{(k)}\}$ for kth powers.

DEFINITION 1.1. A square non-negative matrix T is said to be *primitive* if there exists a positive integer k such that $T^k > 0$.

It is clear that if any other matrix $\tilde{T}$ has the same dimensions as T, and has positive entries and zero entries in the same positions as T, then this will also be true of all powers T^k, $\tilde{T}^k$ of the two matrices.

As *incidence matrix* $\tilde{T}$ corresponding to a given T replaces all the positive entries of T by ones. Clearly $\tilde{T}$ is primitive if and only if T is.

1.1 The Perron–Frobenius Theorem for primitive matrices†

THEOREM 1.1. Suppose T is an $n \times n$ non-negative primitive matrix. Then there exists an eigenvalue r such that:

(a) r real, > 0;
(b) with r can be associated strictly positive left and right eigenvectors;

† This theorem is fundamental to the entire book. The proof is necessarily long; the reader may wish to defer detailed consideration of it.

(*c*) $r > |\lambda|$ for any eigenvalue $\lambda \neq r$;
(*d*) the eigenvectors associated with r are unique to constant multiples;
(*e*) If $0 \leqslant B \leqslant T$ and β is an eigenvalue of B, then $|\beta| \leqslant r$. Moreover, $|\beta| = r$ implies $B = T$.
(*f*) r is a simple root of the characteristic equation of T.

Proof. (*a*) Consider initially a row vector $\boldsymbol{x}' \geqslant \boldsymbol{0}', \neq \boldsymbol{0}'$; and the product $\boldsymbol{x}'\,T$. Let

$$r(\boldsymbol{x}) = \min_j \frac{\sum_i x_i t_{ij}}{x_j}$$

where the ratio is to be interpreted as '∞' if $x_j = 0$. Clearly, $0 \leqslant r(\boldsymbol{x}) < \infty$. Now since

$$x_j\, r(\boldsymbol{x}) \leqslant \sum_i x_i t_{ij} \text{ for each } j,$$

$$\boldsymbol{x}'\, r(\boldsymbol{x}) \leqslant \boldsymbol{x}'\,T,$$

and so

$$\boldsymbol{x}'\boldsymbol{1}\, r(\boldsymbol{x}) \leqslant \boldsymbol{x}'\,T\boldsymbol{1}.$$

Since $T\boldsymbol{1} \leqslant K\boldsymbol{1}$ for $K = \max_i \sum_j t_{ij}$, it follows that

$$r(\boldsymbol{x}) \leqslant \boldsymbol{x}'\,\boldsymbol{1}\,K/\boldsymbol{x}'\boldsymbol{1} = K = \max_i \sum_j t_{ij}$$

so $r(\boldsymbol{x})$ is uniformly bounded above for all such $\boldsymbol{x}$. We note also that since T, being primitive, can have no column consisting entirely of zeroes, $r(\boldsymbol{1}) > 0$, whence it follows that

$$r = \sup_{\substack{\boldsymbol{x} \geqslant \boldsymbol{0} \\ \boldsymbol{x} \neq \boldsymbol{0}}} \min_j \frac{\sum_i x_i t_{ij}}{x_j} \tag{1.1}$$

satisfies

$$0 < r(\boldsymbol{1}) \leqslant r \leqslant K < \infty.$$

Moreover, since neither numerator or denominator is altered by the norming of $\boldsymbol{x}$,

$$r = \sup_{\substack{\boldsymbol{x} \geqslant \boldsymbol{0} \\ \boldsymbol{x}'\boldsymbol{x} = 1}} \min_j \frac{\sum_i x_i t_{ij}}{x_j}.$$

Now the region $\{\boldsymbol{x};\, \boldsymbol{x} \geqslant \boldsymbol{0}, \boldsymbol{x}'\boldsymbol{x} = 1\}$ is compact in the Euclidean n-space R_n, and the function $r(\boldsymbol{x})$ is an upper-semicontinuous mapping of this region into R_1; hence† the supremum, r is actually *attained* for some $\boldsymbol{x}$, say $\hat{\boldsymbol{x}}$. Thus there exists $\hat{\boldsymbol{x}} \geqslant \boldsymbol{0}, \neq \boldsymbol{0}$ such that

$$\min_j \frac{\sum_i \hat{x}_i t_{ij}}{\hat{x}_j} = r,$$

i.e.

$$\sum_i \hat{x}_i t_{ij} \geqslant r\,\hat{x}_j; \quad \text{or} \quad \hat{\boldsymbol{x}}'\,T \geqslant r\,\hat{\boldsymbol{x}}' \tag{1.2}$$

† see Appendix D, §D.1

for each $j = 1, \ldots, n$; with equality for some element of $\hat{\boldsymbol{x}}$.

Now consider

$$\boldsymbol{z}' = \hat{\boldsymbol{x}}' T - r\hat{\boldsymbol{x}}' \geqslant \boldsymbol{0}'.$$

Either $\boldsymbol{z} = \boldsymbol{0}$, or not; if not, we know that for $k \geqslant k_0$, $T^k > 0$ as a consequence of the primitivity of T, and so

$$\boldsymbol{z}' T^k = (\hat{\boldsymbol{x}}' T^k)T - r(\hat{\boldsymbol{x}}' T^k) > \boldsymbol{0}',$$

i.e.
$$\frac{\{(\hat{\boldsymbol{x}}' T^k)T\}_j}{\{\hat{\boldsymbol{x}}' T^k\}_j} > r, \text{ each } j,$$

where the subscript j refers to the jth element. This is a contradiction to the definition of r. Hence always

$$\boldsymbol{z} = \boldsymbol{0},$$

whence
$$\hat{\boldsymbol{x}}' T = r\hat{\boldsymbol{x}}' \tag{1.3}$$

which proves (a).

(b) By iterating (1.3)

$$\hat{\boldsymbol{x}}' T^k = r^k \hat{\boldsymbol{x}}'$$

and taking k sufficiently large $T^k > 0$, and since $\hat{\boldsymbol{x}} \geqslant \boldsymbol{0}, \neq \boldsymbol{0}$, in fact $\hat{\boldsymbol{x}}' > \boldsymbol{0}'$.

(c) Let λ be any eigenvalue of T. Then for some $\boldsymbol{x} \neq \boldsymbol{0}$ and possibly complex valued

$$\sum_i x_i t_{ij} = \lambda x_j \qquad \left(\text{so that } \sum_i x_i t_{ij}^{(k)} = \lambda^k x_j\right) \tag{1.4}$$

whence
$$|\lambda x_j| = \left|\sum_i x_i t_{ij}\right| \leqslant \sum_i |x_i| t_{ij},$$

so that
$$|\lambda| \leqslant \frac{\sum_i |x_i| t_{ij}}{|x_j|}$$

where the right side is to be interpreted as '∞' for any $x_j = 0$. Thus

$$|\lambda| \leqslant \min_j \frac{\sum_i |x_i| t_{ij}}{|x_j|},$$

and by the definition (1.1) of r

$$|\lambda| \leqslant r.$$

Now suppose $|\lambda| = r$; then

$$\sum_i |x_i| t_{ij} \geqslant |\lambda|\, |x_j| = r|x_j|$$

which is a situation identical to that in the proof of part (a), (1.2); so that eventually in the same way

$$\sum_i |x_i| t_{ij} = r|x_j|, > 0; \quad j = 1, 2, \ldots, n \tag{1.5}$$

and so
$$\sum_i |x_i| t_{ij}^{(k)} = r^k |x_j|, > 0; \quad j = 1, 2, \ldots, n,$$

i.e.
$$|\sum_i x_i t_{ij}^{(k)}| = |\lambda^k x_j| = \sum_i |x_i t_{ij}^{(k)}| \tag{1.6}$$

where k can be chosen so large that $T^k > 0$, by the *primitivity* assumption on T; but for two numbers $\gamma, \delta \neq 0$, $|\gamma + \delta| = |\gamma| + |\delta|$ if and only if γ, δ have the same direction in the complex plane. Thus writing $x_j = |x_j| \exp i\theta_j$, (1.6) implies $\theta_j = \theta$ is independent of j, and hence cancelling the exponential throughout (1.4) we get

$$\sum_i |x_i| t_{ij} = \lambda |x_j|$$

where, since $|x_i| > 0$ all i, λ is real and positive, and since we are assuming $|\lambda| = r$, $\lambda = r$ (or the fact follows equivalently from (1.5)).

(*d*) Suppose $\boldsymbol{x}' \neq \boldsymbol{0}'$ is a left eigenvector (possibly with complex elements) corresponding to r.

Then, by the argument in (*c*), so is $\boldsymbol{x}'_+ = \{|x_i|\} \neq \boldsymbol{0}'$, which in fact satisfies $\boldsymbol{x}_+ > \boldsymbol{0}$. Clearly

$$\boldsymbol{\eta}' = \hat{\boldsymbol{x}}' - c\,\boldsymbol{x}'$$

is then also a left eigenvector corresponding to r, for any c such that $\boldsymbol{\eta} \neq \boldsymbol{0}$; and hence the same things can be said about $\boldsymbol{\eta}$ as about $\boldsymbol{x}$; in particular $\boldsymbol{\eta}_+ > \boldsymbol{0}$.

Now either $\boldsymbol{x}$ is a multiple of $\hat{\boldsymbol{x}}$ or not; if not c can be chosen so that $\boldsymbol{\eta} \neq \boldsymbol{0}$, but some element of $\boldsymbol{\eta}$ is; this is impossible as $\boldsymbol{\eta}_+ > \boldsymbol{0}$.

Hence $\boldsymbol{x}'$ is a multiple of $\hat{\boldsymbol{x}}'$.

Right eigenvectors. The arguments (*a*)–(*d*) can be repeated separately for right eigenvectors; (*c*) guarantees that the r produced is the same, since it is purely a statement about eigenvalues.

(*e*) Let $\boldsymbol{y} \neq \boldsymbol{0}$ be a right eigenvector of B corresponding to β. Then taking moduli as before

$$|\beta|\, \boldsymbol{y}_+ \leqslant B\, \boldsymbol{y}_+, \leqslant T\, \boldsymbol{y}_+, \tag{1.7}$$

so that using the same $\hat{\boldsymbol{x}}$ as before

$$|\beta|\, \hat{\boldsymbol{x}}' \boldsymbol{y}_+ \leqslant \hat{\boldsymbol{x}}'\, T\boldsymbol{y}_+ = r\, \hat{\boldsymbol{x}}'\, \boldsymbol{y}_+$$

and since $\hat{\boldsymbol{x}}'\, \boldsymbol{y}_+ > 0$,

$$|\beta| \leqslant r.$$

Suppose now $|\beta| = r$; then from (1.7)

$$r\,y_+ \leqslant T\,y_+$$

whence, as in the proof of (*b*), it follows $T\,y_+ = r\,y_+ > \mathbf{0}$; whence from (1.7)

$$r\,y_+ = B\,y_+ = T\,y_+$$

so it must follow, from $B \leqslant T$, that $B = T$.

(*f*) The following identities are true for all numbers, real and complex, including eigenvalues of T:

$$\left.\begin{aligned}(xI - T)\,\mathrm{Adj}\,(xI - T) &= \det\,(xI - T)I\\ \mathrm{Adj}\,(xI - T)\,(xI - T) &= \det\,(xI - T)I\end{aligned}\right\} \tag{1.8}$$

where I is the unit matrix and 'det' refers to the determinant. (The relation is clear for x not an eigenvalue, since then $\det(xI - T) \neq 0$; when x is an eigenvalue it follows by continuity.)

Put $x = r$: then any one row of $\mathrm{Adj}(rI - T)$ is either (i) a left eigenvector corresponding to r; or (ii) a row of zeroes; and similarly for columns. By assertions (*b*) and (*d*) (already proved) of the theorem, $\mathrm{Adj}\,(rI - T)$ is either (i) a matrix with no elements zero; or (ii) a matrix with all elements zero. We shall prove that one element of $\mathrm{Adj}\,(rI - T)$ is positive, which establishes that case (i) holds. The (n,n) element is

$$\det\,(r\;{}_{(n-1)}I - {}_{(n-1)}T)$$

where ${}_{(n-1)}T$ is T with last row and column deleted; and ${}_{(n-1)}I$ is the corresponding unit matrix. Since

$$0 \leqslant \begin{bmatrix} {}_{(n-1)}T & \mathbf{0} \\ \mathbf{0}' & 0 \end{bmatrix} \leqslant T, \text{ and } \neq T,$$

the last since T is primitive (and so can have no zero column), it follows from (*e*) of the theorem that no eigenvalue of ${}_{(n-1)}T$ can be as great in modulus as r. Hence

$$\det\,(r\;{}_{(n-1)}I - {}_{(n-1)}T) > 0,$$

as required; *and moreover we deduce that* $\mathrm{Adj}\,(rI - T)$ *has all its elements positive.*

Write $\phi(x) = \det\,(xI - T)$; then differentiating (1.8) elementwise

$$\mathrm{Adj}\,(xI - T) + (xI - T)\,\frac{\mathrm{d}}{\mathrm{d}x}\,\{\mathrm{Adj}\,(xI - T)\} = \phi'\,(x)I.$$

Substitute $x = r$, and premultiply by $\hat{x}'$;

$$(0' <)\ \hat{x}'\ \mathrm{Adj}\,(rI - T) = \phi'(r)\,\hat{x}'$$

since the other term vanishes. Hence $\phi'(r) > 0$ and so r is simple. #

COROLLARY 1.

$$\min_i \sum_{j=1}^{n} t_{ij} \leqslant r \leqslant \max_i \sum_{j=1}^{n} t_{ij} \tag{1.9}$$

with equality on either side implying equality throughout (i.e. r can only be equal to the maximal or minimal row sum if all row sums are equal).

A similar proposition holds for column sums.

Proof. Recall from the proof of part (*a*) of the theorem, that

$$0 < r(\mathbf{1}) = \min_j \sum_i t_{ij} \leqslant r \leqslant K = \max_i \sum_j t_{ij} < \infty. \tag{1.10}$$

Since T' is also primitive and has the same r, we have also

$$\min_j \sum_i t_{ji} \leqslant r \leqslant \max_i \sum_j t_{ji} \tag{1.11}$$

and a combination of (1.10) and (1.11) gives (1.9).

Now assume that one of the equalities in (1.9) holds, but not all row sums are equal. Then by increasing (or, if appropriate, decreasing) the positive elements of T (but keeping them positive), produce a new primitive matrix, with all row sums equal and the same r, in view of (1.9); which is impossible by assertion (*e*) of the theorem. #

COROLLARY 2. Let $\boldsymbol{v}'$ and $\boldsymbol{w}$ be positive left and right eigenvectors corresponding to r, normed so that $\boldsymbol{v}'\,\boldsymbol{w} = 1$. Then

$$\mathrm{Adj}\,(rI - T)\,/\phi'(r) = \boldsymbol{w}\,\boldsymbol{v}'.$$

To see this, first note that since the columns of Adj $(rI - T)$ are multiples of the same positive right eigenvector corresponding to r (and its rows of the same positive left eigenvector) it follows that we can write it in the form $\boldsymbol{y}\,\boldsymbol{x}'$ where $\boldsymbol{y}$ is a right and $\boldsymbol{x}'$ a left positive eigenvector. Moreover, again by uniqueness, there exist positive constants c_1, c_2 such that $\boldsymbol{y} = c_1\,\boldsymbol{w}$, $\boldsymbol{x}' = c_2\,\boldsymbol{v}'$, whence

$$\mathrm{Adj}\,(rI - T) = c_1\,c_2\,\boldsymbol{w}\,\boldsymbol{v}'.$$

Now, as in the proof of the simplicity of r,

$$\boldsymbol{v}'\,\phi'(r) = \boldsymbol{v}'\ \mathrm{Adj}\,(rI - T) = c_1\,c_2\,\boldsymbol{v}'\,\boldsymbol{w}\,\boldsymbol{v}' = c_1\,c_2\,\boldsymbol{v}'$$

so that $\boldsymbol{v}'\,\boldsymbol{w}\,\phi'(r) = c_1\,c_2\,\boldsymbol{v}'\,\boldsymbol{w}$

i.e. $\quad c_1\,c_2 = \phi'(r)$ as required. #

(Note that $c_1\,c_2$ = sum of the diagonal elements of the adjoint = sum of the principal $(n-1) \times (n-1)$ minors of $(rI - T)$.)

Theorem 1.1 is the strong version of the Perron–Frobenius Theorem which holds for primitive T; we shall generalize Theorem 1.1 to a wider class of matrices, called *irreducible* in §1.4 (and shall refer to this generalization as the Perron–Frobenius Theory).

Suppose now the *distinct* eigenvalues of a primitive T are $r, \lambda_2, \ldots, \lambda_t$, $t \leqslant n$ where $r > |\lambda_2| \geqslant |\lambda_3| \geqslant \ldots \geqslant |\lambda_t|$. In the case $|\lambda_2| = |\lambda_3|$ we stipulate that the multiplicity m_2 of λ_2 is at least as great as that of λ_3, *and of any other eigenvalue having the same modulus as* λ_2.

THEOREM 1.2. For a primitive matrix T, as $k \to \infty$

$$T^k = r^k \, \boldsymbol{w} \, \boldsymbol{v}' + 0(k^s |\lambda_2|^k)$$

elementwise, where $\boldsymbol{w}$, $\boldsymbol{v}'$ are any positive right and left eigenvectors corresponding to r guaranteed by Theorem 1.1, providing only they are normed so that $\boldsymbol{v}' \boldsymbol{w} = 1$. Here $s = m_2 - 1$.

Proof. Let $R(z) = (I - zT)^{-1}$

$= \{r_{ij}(z)\}$, $z \neq \lambda_i^{-1}$, $i = 1, 2, \ldots$ (where $\lambda_1 = r$). Consider a general element of this matrix:

$$r_{ij}(z) = \frac{c_{ij}(z)}{(1 - zr)(1 - z\lambda_2)^{m_2} \ldots (1 - z\lambda_t)^{m_t}}$$

where m_i is the multiplicity of λ_i and $c_{ij}(z)$ a polynomial in z, of degree at most $n-1$ (see Appendix B). Hence by partial fractions

$$r_{ij}(z) = \frac{a_{ij}}{(1 - zr)} + \frac{\sum_{i=0}^{m_2-1} b_{ij}^{(m_2-i)}}{(1 - z\lambda_2)^{m_2-i}} + \text{similar terms for other eigenvalues,}$$

where the a_{ij}, $b_{ij}^{(m_2-i)}$ are constants.

Hence for $|z| < 1/r$,

$$r_{ij}(z) = a_{ij} \sum_{k=0}^{\infty} (zr)^k + \sum_{i=0}^{m_2-1} b_{ij}^{(m_2-i)} \left\{ \sum_{k=0}^{\infty} \begin{bmatrix} -m_2 + i \\ k \end{bmatrix} (-z\lambda_2)^k \right\} + \begin{array}{c}\text{similar terms}\\ \text{for other}\\ \text{eigenvalues;}\end{array}$$

i.e. elementwise it is true

$$R(z) = A \sum_{k=0}^{\infty} (zr)^k + \sum_{i=0}^{m_2-1} B^{(m_2-i)} \left\{ \sum_{k=0}^{\infty} \begin{bmatrix} -m_2 + i \\ k \end{bmatrix} (-z\lambda_2)^k \right\} + \text{like terms,}$$

where $A = \{a_{ij}\}$ etc.

From Stirling's formula, as $k \to \infty$

$\begin{bmatrix} m_2 + i \\ k \end{bmatrix} \sim \text{const.} k^{m_2-i-1}$, so that, identifying coefficients of z^k on both sides (see Appendix B) for large k

$$T^k = Ar^k + 0(k^{m_2-1} |\lambda_2|^k).$$

It remains to determine the nature of A. We first note that

$$T^k/r^k \to A \geqslant 0 \text{ elementwise, as } k \to \infty,$$

and that the series

$$\sum_{k=0}^{\infty} (r^{-1}T)^k z^k$$

has non-negative coefficients, and is convergent for $|z| < 1$, so that by a well-known result (see e.g. Heathcote, 1971, p. 65).

$$\lim_{x \to 1-} (1 - x) \sum_{k=0}^{\infty} (r^{-1}T)^k x^k = A \text{ elementwise.}$$

Now, for $0 < x < 1$,

$$\sum_{k=0}^{\infty} (r^{-1}T)^k x^k = (I - r^{-1}xT)^{-1} = \frac{\text{Adj}\,(I - r^{-1}xT)}{\det\,(I - r^{-1}xT)}$$

$$= \frac{r}{x} \frac{\text{Adj}\,(rx^{-1}I - T)}{\det\,(rx^{-1}I - T)}$$

so that $\quad A = -r\,\text{Adj}\,(rI - T)/c$

where $\quad c = \lim_{x \to 1-} \{-\det\,(rx^{-1}I - T)/(1 - x)\}$

$$= \frac{\mathrm{d}}{\mathrm{d}x}[\phi(rx^{-1})]_{x=1}$$

$$= -r\,\phi'(r)$$

which completes the proof, taking into account Corollary 2 of Theorem 1.1 #

In conclusion to this section we point out that assertion (*d*) of Theorem 1.1 states that the *geometric multiplicity* of the eigenvalue r is one, whereas (*f*) states that its *algebraic multiplicity* is one. It is well known in matrix theory that geometric multiplicity one for the eigenvalue of a square arbitrary matrix does not in general imply algebraic multiplicity one. A simple example to this end is the matrix (which is non-negative, but of course not primitive):

$$\begin{bmatrix} 1 & 1 \\ 0 & 1 \end{bmatrix}$$

which has repeated eigenvalue unity (algebraic multiplicity two), but a corresponding left eigenvector can only be a multiple of $\{0,1\}$ (geometric multiplicity one).

The distinction between geometric and algebraic multiplicity in connection with r in a primitive matrix is slurred over in some treatments of non-negative matrix theory.

1.2 Structure of a general non-negative matrix

In this section we are concerned with a general square matrix $T = \{t_{ij}\}$, $i, j = 1, \ldots, n$, satisfying $t_{ij} \geqslant 0$, with the aim of showing that the behaviour of its powers T^k reduces, to a substantial extent, to the behaviour of powers of a fundamental type of non-negative square matrix, called *irreducible.* The class of irreducible matrices further subdivides into matrices which are *primitive* (studied in §1.1), and *cyclic* (imprimitive), whose study is taken up in §1.3.

We introduce here a definition, which, while frequently used in other expositions of the theory, and so possibly useful to the reader, will be used by us only to a limited extent.

DEFINITION 1.2. A sequence $(i, i_1, i_2, \ldots, i_{t-1}, j)$, for $t \geqslant 1$ (where $i_0 = i$), from the index set $\{1, 2, \ldots, n\}$ of a non-negative matrix T is said to form a *chain* of length t between the ordered pair (i,j) if

$$t_{ii_1} t_{i_1 i_2} \cdots t_{i_{t-2} i_{t-1}} t_{i_{t-1} j} > 0.$$

Such a chain for which $i = j$ is called a *cycle* of length t between i and itself.

Clearly in this definition, we may without loss of generality impose the restriction that, for fixed (i, j), $i, j \neq i_1 \neq i_2 \neq \ldots \neq i_{t-1}$, to obtain a 'minimal' length chain or cycle, from a given one.

Classification of indices

Let i, j, k be arbitrary indices from the index set $1, 2, \ldots, n$ of the matrix T.

We say that *i leads to j,* and write $i \to j$, if there exists an integer $m \geqslant 1$ such that $t_{ij}^{(m)} > 0$.† If i does not lead to j we write $i \not\to j$. Clearly, if $i \to j$ and $j \to k$ then, from the rule of matrix multiplication, $i \to k$.

We say that i and j *communicate* if $i \to j$ and $j \to i$, and write in this case $i \leftrightarrow j$.

The indices of the matrix T may then be classified and grouped as follows.

(*a*) If $i \to j$ but $j \not\to i$ for some j, then the index i is called *inessential.* An index which leads to no index at all (this arises when T has a row of zeros) is also called inessential.

(*b*) Otherwise the index i is called *essential*. Thus if i is essential, $i \to j$ implies $i \leftrightarrow j$; and there is at least one j such that $i \to j$.

(*c*) It is therefore clear that all essential indices (if any) can be subdivided into *essential classes* in such a way that all the indices belonging to one class communicate, but cannot lead to an index outside the class.

(*d*) Moreover, all inessential indices (if any) which communicate with some index, may be divided into *inessential classes* such that all indices in a class communicate.

† Or, equivalently, if there is a chain between i and j.

Classes of the type described in (*c*) and (*d*) are called *self-communicating* classes.

(*e*) In addition there may be inessential indices which communicate with no index; these are defined as forming an *inessential class* by themselves (which, of course, is not self-communicating). These are of nuisance value only as regards applications, but are included in the description for completeness.

This description appears complex, but should be much clarified by the example which follows, and similar exercises.

Before proceeding, we need to note that the classification of indices (and hence grouping into classes) *depends only on the location of the positive elements, and not on their size,* so any two non-negative matrices with the same incidence matrix will have the same index classification and grouping (and, indeed, *canonical form,* to be discussed shortly).

Further, given a non-negative matrix (or its incidence matrix), classification and grouping of indices is made easy by a *path diagram* which may be described as follows. Start with index 1 – this is the zeroth stage; determine all j such that $1 \to j$ and draw arrows to them – these j form the 2nd stage; for each of these j now repeat the procedure to form the 3rd stage; and so on; *but* as soon as an index occurs which has occurred at an earlier stage, ignore further consequents of it. Thus the diagram terminates when every index in it has repeated. (since there are a finite total number of indices, the process must terminate.) This diagram will represent all possible consequent behaviour for the set of indices which entered into it, which may not, however be the entire index set. If any are left over, choose one such and draw a similar diagram for it, regarding the indices of the previous diagram also as having occurred 'at an earlier stage'. And so on, till all indices of the index set are accounted for.

Example. A non-negative matrix T has incidence matrix

$$\begin{array}{c|ccccccccc|} & 1 & 2 & 3 & 4 & 5 & 6 & 7 & 8 & 9 \\ \hline 1 & 1 & 1 & 0 & 0 & 0 & 0 & 0 & 0 & 0 \\ 2 & 1 & 1 & 1 & 0 & 0 & 0 & 1 & 0 & 0 \\ 3 & 0 & 0 & 0 & 0 & 0 & 0 & 1 & 0 & 0 \\ 4 & 0 & 0 & 0 & 1 & 0 & 0 & 0 & 0 & 1 \\ 5 & 0 & 0 & 0 & 0 & 1 & 0 & 0 & 0 & 0 \\ 6 & 0 & 0 & 1 & 0 & 0 & 1 & 0 & 0 & 0 \\ 7 & 0 & 0 & 1 & 0 & 0 & 0 & 0 & 0 & 0 \\ 8 & 0 & 1 & 0 & 0 & 0 & 1 & 0 & 1 & 0 \\ 9 & 0 & 0 & 0 & 1 & 0 & 0 & 0 & 0 & 1 \end{array}$$

Thus Diagram 1 tells us $\{3, 7\}$ is an essential class, while $\{1, 2\}$ is an inessential (communicating) class. (See over.)
Diagram 2 tells us $\{4, 9\}$ is an essential class.
Diagram 3 tells us $\{5\}$ is an essential class.

Diagram 1

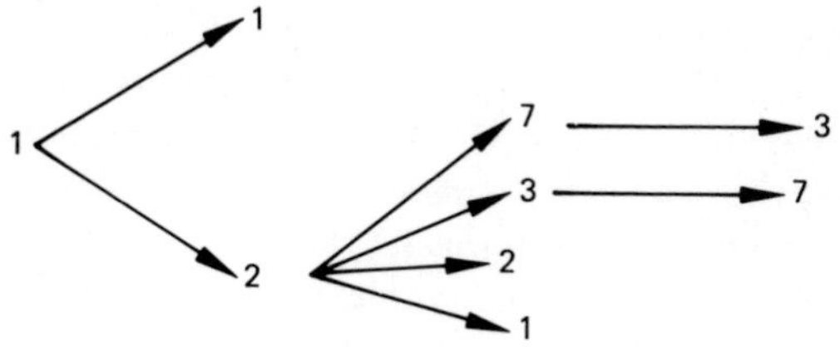

Diagram 2

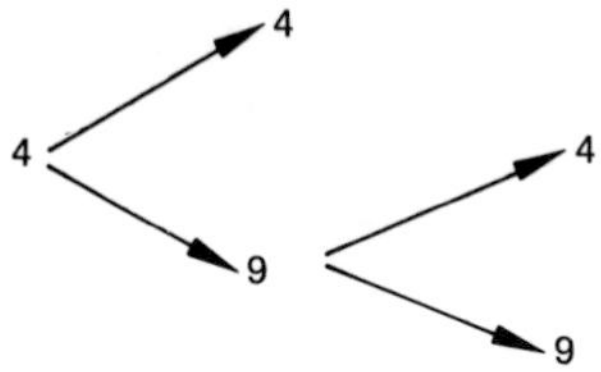

Diagram 3

Diagram 4

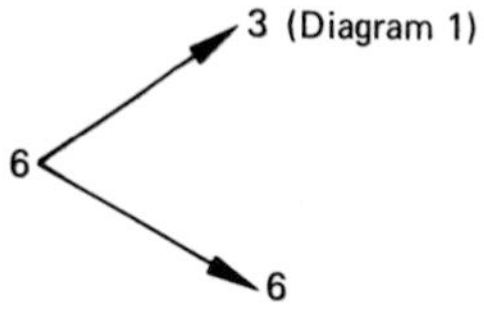

Diagram 5

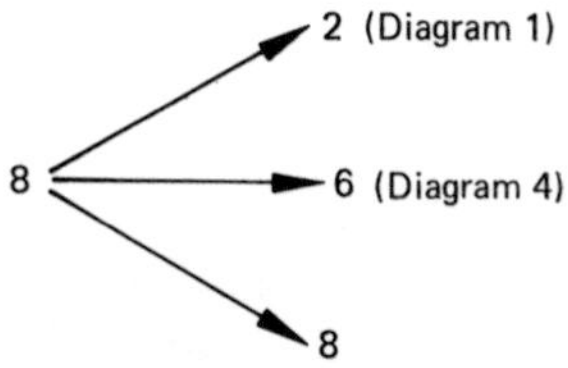

Diagram 4 tells us $\{6\}$ is an inessential (self-communicating) class.
Diagram 5 tells us $\{8\}$ is an inessential (self-communicating) class.

Canonical form

Once the classification and grouping has been carried out, the definition 'leads' may be extended to classes in the obvious sense e.g. the statement $\mathscr{C}_1 \to \mathscr{C}_2$ ($\mathscr{C}_1 \neq \mathscr{C}_2$) means that there is an index of $\mathscr{C}_1$ which leads to an index of $\mathscr{C}_2$. Hence all indices of $\mathscr{C}_1$ lead to all indices of $\mathscr{C}_2$, and the statement can only apply to an inessential class $\mathscr{C}_1$.

Moreover, the matrix T may be put into *canonical form* by first relabelling the indices in a specific manner. Before describing the manner, we mention that relabelling the indices using the same index set $\{1, \ldots, n\}$ and rewriting T accordingly merely amounts to performing a *simultaneous permutation* of rows and columns of the matrix. Now such a simultaneous permutation only amounts to a *similarity transformation* of the original matrix, T, so that (*a*) its powers are similarly transformed; (*b*) its spectrum (i.e. set of eigenvalues) is unchanged. Generally any given ordering is as good as any other in a physical context; but the canonical form of T, arrived at by one such ordering, is particularly convenient.

The canonical form is attained by first taking the indices of one essential class (if any) and renumbering them consecutively using the lowest integers, and following by the indices of another essential class, if any, until the essential classes are exhausted. The numbering is then extended to the indices of the inessential classes (if any) which are themselves arranged in an order such that an inessential class occurring earlier (and thus higher in the arrangement) does not *lead* to any inessential class occurring later.

Example (continued): For the given matrix T the essential classes are $\{5\}$, $\{4, 9\}$, $\{3, 7\}$; and the inessential classes $\{1, 2\}$, $\{6\}$, $\{8\}$ which from Diagrams 4 and 5 should be ranked in this order. Thus a possible canonical form for T is

	5	4	9	3	7	1	2	6	8
5	1	0	0	0	0	0	0	0	0
4	0	1	1	0	0	0	0	0	0
9	0	1	1	0	0	0	0	0	0
3	0	0	0	0	1	0	0	0	0
7	0	0	0	1	0	0	0	0	0
1	0	0	0	0	0	1	1	0	0
2	0	0	0	1	1	1	1	0	0
6	0	0	0	1	0	0	0	1	0
8	0	0	0	0	0	0	1	1	1

It is clear that the canonical form consists of square diagonal blocks corresponding to 'transition within' the classes in one 'stage', zeros to the right of these diagonal blocks, but at least one non-zero element to the left of each inessential block unless it corresponds to an index which leads to no other. Thus the general version of the canonical form of T is

$$T = \left[\begin{array}{cccc|c} T_1 & 0 & 0 \ldots & & 0 \\ 0 & T_2 & & & \cdot \\ & 0 & & & \cdot \\ \cdot & & & & \cdot \\ \cdot & & & & \\ \cdot & & & & \\ 0 & 0 & \ldots & T_z & 0 \\ \hline & R & & & Q \end{array}\right]$$

where the T_i correspond to the z essential classes, and Q to the inessential indices, with $R \neq 0$ in general, with Q itself having a structure analogous to T, except that there may be non-zero elements to the left of any of its diagonal blocks:

$$Q = \begin{bmatrix} Q_1 & & & & \\ & Q_2 & & & \\ & & \cdot & & \\ & & & \cdot & 0 \\ & & & & \cdot \\ S & & & & Q_w \end{bmatrix}.$$

Now, in most applications we are interested in the behaviour of the *powers* of T. Let us assume it is in canonical form. Since

$$T^k = \left[\begin{array}{ccccc|c} T_1^k & & & & & \\ & T_2^k & & & & \\ & & \cdot & & & \\ 0 & & & \cdot & 0 & 0 \\ & & & & \cdot & \\ & & & & T_z^k & \\ \hline & & R_k & & & Q^k \end{array}\right], \quad Q^k = \begin{bmatrix} Q_1^k & & & & \\ & Q_2^k & & & \\ & & \cdot & & \\ & & & \cdot & 0 \\ & & & & \cdot \\ S_k & & & & Q_w^k \end{bmatrix}$$

it follows that a substantial advance in this direction will be made in studying the powers of the diagonal *block submatrices corresponding to self–communicating classes* (the other diagonal block submatrices, if any, are 1 x 1 zero matrices; the evolution of R_k and S_k is complex, with k). In fact if one is interested in only the essential indices, as is often the case, this is sufficient.

A (sub)matrix corresponding to a single self-communicating class is called *irreducible.*

It remains to show that, *normally,* there is at least one self-communicating (indeed essential) class of indices present for any matrix T; although it is nevertheless *possible* that all indices of a non-negative matrix fall into non self-communicating classes (and are therefore inessential): for example

$$T = \begin{bmatrix} 0 & 0 \\ 1 & 0 \end{bmatrix}$$

LEMMA 1.1. An $n \times n$ non-negative matrix with at least one positive entry in each row possesses at least one essential class of indices.

Proof. Suppose all indices are inessential. The assumption of non-zero rows then implies that for any index i, $i = 1, \ldots, n$, there is at least one j such that $i \to j$, but $j \not\to i$.

Now suppose i_1 is any index. Then we can find a sequence of indices $i_2, i_3, \ldots$ etc such that

$$i_1 \to i_2 \to i_3 \to \ldots \to i_n \to i_{n+1} \ldots$$

but such that $i_{k+1} \nrightarrow i_k$, and hence $i_{k+1} \nrightarrow i_1, i_2, \ldots,$ or i_{k-1}. However since the sequence $i_1, i_2, \ldots, i_{n+1}$ is a set of $n + 1$ indices, each chosen from the same n possibilities, $1, 2, \ldots, n$, at least *one index repeats* in the sequence. This is a contradiction to the deduction that no index can lead to an index with a lower subscript. #

We come now to the important concept of the period of an index.

DEFINITION 1.3. If $i \rightarrow i$, $d(i)$ is the *period* of the index i if it is the greatest common divisor of those k for which

$$t_{ii}^{(k)} > 0$$

(see Definition A.2 in Appendix A). N.B. If $t_{ii} > 0$, $d(\mathrm{i}) = 1$.

We shall now prove that in a communicating class all indices have the same period.

LEMMA 1.2. If $i \leftrightarrow j$, $d(i) = d(j)$.

Proof. Let $t_{ij}^{(M)} > 0$, $t_{ji}^{(N)} > 0$. Then for any positive integer s such that $t_{jj}^{(s)} > 0$.

$$t_{ii}^{(M+s+N)} \geqslant t_{ij}^{(M)}\, t_{jj}^{(s)}\, t_{ji}^{(N)} > 0,$$

the first inequality following from the rule of matrix multiplication and the non-negativity of the elements of T. Now, for such an s it is also true that $t_{jj}^{(2s)} > 0$ necessarily, so that

$$t_{ii}^{(M+2s+N)} > 0.$$

Therefore $\quad d(i)$ divides $M + 2s + N - (M + s + N) = s$.

Hence: for every s such that $t_{jj}^{(s)} > 0$, $d(i)$ divides s.

Hence $$d(i) \leqslant d(j).$$

But since the argument can be repeated with i and j interchanged,

$$d(j) \leqslant d(i).$$

Hence $d(i) = d(j)$ as required. #

Note that, again, consideration of an incidence matrix is adequate to determine that period.

DEFINITION 1.4. The period of a communicating class is the period of any one of its indices.

Example (continued): Determine the periods of all communicating classes for the matrix T with incidence matrix considered earlier.

Essential classes:

$\{5\}$ has period 1, since $t_{55} > 0$.

$\{4,9\}$ has period 1, since $t_{44} > 0$.

$\{3,7\}$ has period 2, since $t_{33}^{(k)} > 0$

for every even k, and is zero for every odd k.

Inessential self-communicating classes:

$\{1,2\}$ has period 1 since $t_{11} > 0$.

$\{6\}$ has period 1 since $t_{66} > 0$.

$\{8\}$ has period 1 since $t_{88} > 0$.

DEFINITION 1.5. An index i such that $i \to i$ is *aperiodic* (acyclic) if $d(i) = 1$. (It is thus contained in an *aperiodic (self-communicating) class.*)

1.3 Irreducible matrices

Towards the end of the last section we called a non-negative square matrix, corresponding to a single self-communicating class of indices, irreducible. We now give a general definition, independent of the previous context, which is, nevertheless, easily seen to be equivalent to the one just given. The part of the definition referring to periodicity is justified by Lemma 1.2.

DEFINITION 1.6. An $n \times n$ non-negative matrix T is *irreducible* if for every pair i, j of its index set, there exists a positive integer $m \equiv m(i, j)$ such that $t_{ij}^{(m)} > 0$. An irreducible matrix is said to be cyclic (periodic) with period d, if the period of any one (and so of each one) of its indices satisfies $d > 1$, and is said to be acyclic (aperiodic) if $d = 1$.

The following results all refer to an irreducible matrix with period d.

Note that an irreducible matrix T cannot have a zero row or column.

LEMMA 1.3. If $i \to i$, $t_{ii}^{(kd)} > 0$ for all integers $k \geq N_0$ $(= N_0(i))$.

Proof.

Suppose $$t_{ii}^{(kd)} > 0,\ t_{ii}^{(sd)} > 0.$$

Then $$t_{ii}^{([k+s]d)} \geq t_{ii}^{(kd)}\, t_{ii}^{(sd)} > 0.$$

Hence the positive integers $\{kd\}$ such that

$$t_{ii}^{(kd)} > 0,$$

form a closed set under addition, and their greatest common divisor is d. An appeal to Lemma A.3 of Appendix A completes the proof. #

THEOREM 1.3. Let i be any fixed index of the index set $\{1, 2, \ldots, n\}$ of T. Then, for evey index j there exists a unique integer r_j in the range $0 \leq r_j < d$

(r_j is called a residue class modulo d) such that

(a) $t_{ij}^{(s)} > 0$ implies $s \equiv r_j \pmod d$;† and
(b) $t_{ij}^{(kd+r_j)} > 0$ for $k \geqslant N(j)$, where $N(j)$ is same positive integer.

Proof. Let $t_{ij}^{(m)} > 0$ and $t_{ij}^{(m')} > 0$.

There exists a p such that $t_{ji}^{(p)} > 0$, whence as before

$$t_{ii}^{(m+p)} > 0 \text{ and } t_{ii}^{(m'+p)} > 0.$$

Hence d divides each of the superscripts, and hence their difference $m-m'$. Thus $m-m' \equiv 0 \pmod d$, so that

$$m \equiv r_j \pmod d.$$

This proves (a).

To prove (b), since $i \to j$ and in view of (a), there exists a positive m such that

$$t_{ij}^{(md+r_j)} > 0.$$

Now, let $N(j) = N_0 + m$,

where N_0 is the number guaranteed by Lemma 1.3 for which $t_{ii}^{(sd)} > 0$ for $s \geqslant N_0$. Hence if $k \geqslant N(j)$, then

$$kd + r_j = sd + md + r_j, \text{ where } s \geqslant N_0.$$

Therefore $t_{ij}^{(kd+r_j)} \geqslant t_{ii}^{(sd)}\, t_{ij}^{(md+r_j)} > 0$, for all $k \geqslant N(j)$. #

DEFINITION 1.7. The set of indices j in $\{1, 2, \ldots, n\}$ corresponding to the same residue class (mod d) is called a subclass of the class $\{1, 2, \ldots, n\}$, and is denoted by C_r, $0 \leqslant r < d$.

It is clear that the d subclasses C_r are disjoint, and their union is $\{1, 2, \ldots, n\}$. It is not yet clear that the composition of the classes does not depend on the choice of initial fixed index i, which we prove in a moment; nor that each subclass contains at least one index.

LEMMA 1.4. The composition of the residue classes does not depend on the initial choice of fixed index i; an initial choice of another index merely subjects the subclasses to a cyclic permutation.

Proof. Suppose we take a new fixed index i'. Then

$$t_{ij}^{(md+r_{i'}+kd+r_j')} \geqslant t_{ii'}^{(kd+r_{i'})}\, t_{i'j}^{(md+r'_j)}$$

where r_j' denotes the residue class corresponding to j according to classification with respect to fixed index i'. Now, by Theorem 1.3 for large k, m, the right hand side is positive, so that the left hand side is also, whence, in the old classification,

$$md + r_{i'} + kd + r_j' \equiv r_j \pmod d$$

i.e.

$$r_{i'} + r_j' \equiv r_j \pmod d.$$

† Recall from Appendix A, that this means that if qd is the multiple of d nearest to s from below, then $s = qd + r_j$; it reads 's is equivalent to r_j, modulo d'.

Hence the composition of the subclasses $\{C_i\}$ is unchanged, and their order is merely subjected to a cyclic permutation σ:

$$\begin{pmatrix} 0 & 1 & \dots & d-1 \\ \sigma(0) & \sigma(1) & \dots & \sigma(d-1) \end{pmatrix}. \quad \#$$

For example, suppose we have a situation with $d = 3$, and $r_{i'} = 2$. Then the classes which were C_0, C_1, C_2 in the old classification according to i (according to which $i' \in C_2$) now become, respectively, C_1', C_2', C_0' since we must have $2 + r_j' \equiv r_j \pmod d$ for $r_j = 0, 1, 2$.

Let us now define C_r for all non-negative integers r by putting $C_r = C_{r_j}$ if $r \equiv r_j \pmod d$, using the initial classification with respect to i. Let m be a positive integer, and consider any j for which $t_{ij}^{(m)} > 0$. (There is at least one appropriate index j, otherwise T^m (and hence higher powers) would have ith row consisting entirely of zeros, contrary to irreducibility of T.) Then $m \equiv r_j \pmod d$, i.e. $m = sd + r_j$ and $j \in C_{r_j}$. Now, similarly, let k by any index such that

$$t_{ik}^{(m+1)} > 0.$$

Then, since $m + 1 = sd + r_j + 1$, it follows $k \in C_{r_j + 1}$.

Hence it follows that, looking at the ith row, the positive entries occur, for successive powers, in successive subclasses. In particular each of the d cyclic classes is non-empty. If subclassification has initially been made according to the index i', since we have seen the subclasses are merely subjected to a cyclic permutation, the classes still 'follow each other' in order, looking at successive powers, and ith (hence any) row.

If follows that if $d > 1$ (so there is more than one subclass) a canonical form of T is possible, by relabelling the indices so that the indices of C_0 come first, of C_1 next, and so on. This produces a version of T of the sort

$$T_c = \begin{bmatrix} 0 & Q_{01} & 0 & \dots & & 0 \\ 0 & 0 & Q_{12} & \dots & & 0 \\ \cdot & \cdot & 0 & & & \\ \cdot & \cdot & \cdot & 0 & & \\ 0 & \cdot & \cdot & \cdot & & Q_{d-2,\,d-1} \\ Q_{d-1,0} & 0 & 0 & \dots & & 0 \end{bmatrix}$$

Example: Check that the matrix, whose incidence matrix is given below is irreducible, find its period, and put into a canonical form if periodic.

	1	2	3	4	5	6
1	0	1	0	0	0	0
2	0	0	1	1	0	1
3	1	0	0	0	1	0
4	1	0	0	0	0	0
5	0	1	0	0	0	0
6	0	0	0	0	1	0

Diagram 1

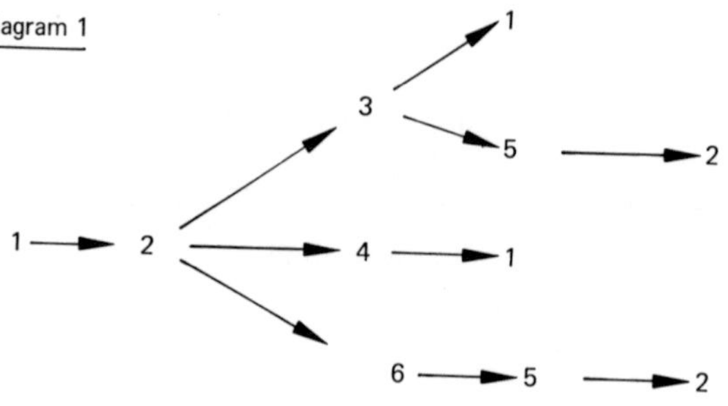

Clearly $i \to j$ for any i and j in the index set, so the matrix is certainly irreducible. Let us now carry out the determination of subclasses on the basis of index 1. Therefore index 1 must be in the subset C_0; 2 must be in C_1; 3,4,6 in C_2; 1,5 in C_3; 2 in C_4. Hence C_0 and C_3 are identical; C_1 and C_4; etc., and so $d = 3$. Moreover

$$C_0 = \{1,5\}, C_1 = \{2\}, C_2 = \{3,4,6\},$$

so canonical form is

	1	5	2	3	4	6
1	0	0	1	0	0	0
5	0	0	1	0	0	0
2	0	0	0	1	1	1
3	1	1	0	0	0	0
4	1	0	0	0	0	0
6	0	1	0	0	0	0

THEOREM 1.4. An irreducible acyclic matrix T is primitive and conversely. The powers of an irreducible cyclic matrix may be studied in terms of powers of primitive matrices.

Proof. If T is irreducible, with $d = 1$, there is only one subclass of the index set, consisting of the index set itself, and Theorem 1.3 implies

$$t_{ij}^{(k)} > 0 \text{ for } k \geqslant N(i, j).$$

Hence for $N^* = \max_{i,j} N(i, j)$

$$t_{ij}^{(k)} > 0, k \geqslant N^*, \text{for all } i, j.$$

i.e.
$$T^k > 0 \text{ for } k \geqslant N^*.$$

Conversely, a primitive matrix is trivially irreducible, and has $d = 1$, since for any fixed i, and k great enough $t_{ii}^{(k)} > 0$, $t_{ii}^{(k+1)} > 0$, and the greatest common divisor of k and $k + 1$ is 1.

The truth of the second part of the assertion may be conveniently demonstrated in the case $d = 3$, where the canonical form of T is

$$T_c = \begin{bmatrix} 0 & Q_{01} & 0 \\ 0 & 0 & Q_{12} \\ Q_{20} & 0 & 0 \end{bmatrix},$$

$$\text{and } T_c^2 = \begin{bmatrix} 0 & 0 & Q_{01}Q_{12} \\ Q_{12}Q_{20} & 0 & 0 \\ 0 & Q_{20}Q_{01} & 0 \end{bmatrix},$$

$$T_c^3 = \begin{bmatrix} Q_{01}Q_{12}Q_{20} & 0 & 0 \\ 0 & Q_{12}Q_{20}Q_{01} & 0 \\ 0 & 0 & Q_{20}Q_{01}Q_{12} \end{bmatrix}.$$

Now, the diagonal matrices of T_c^3 (of T_c^d in general) are square and *primitive,* for Lemma 1.3 states that $t_{ii}^{(3k)} > 0$ for all k sufficiently large. Hence

$$T_c^{3k} = (T_c^3)^k,$$

so that powers which are integral multiples of the period may be studied with the aid of the primitive matrix theory of §1.1. One needs to consider also

$$T_c^{3k+1} \text{ and } T_c^{3k+2}$$

but these present no additional difficulty since we may write $T_c^{3k+1} = (T_c^{3k})T$, $T_c^{3k+2} = (T_c^{3k})T^2$ and proceed as before. #

These remarks substantiate the reason for considering primitive matrices as of prime importance, and for treating them first. It is, nevertheless, convenient to consider a theorem of the type of the fundamental Theorem 1.1 for the broader class of irreducible matrices, which we now expect to be closely related.

1.4 Perron–Frobenius theory for irreducible matrices

THEOREM 1.5. Suppose T is an $n \times n$ irreducible non-negative matrix. Then all of the assertions (a)–(f) of Theorem 1.1 holds, except that (c) is replaced by the weaker statement: $r \geqslant |\lambda|$ for any eigenvalue λ of T. Corollaries 1 and 2 of Theorem 1.1 hold also.

Proof. The proof of (a) of Theorem 1.1 holds to the stage where we need to assume

$$z' = \hat{x}' T - r\,\hat{x}' \geqslant \mathbf{0}' \text{ but } \neq \mathbf{0}'.$$

The matrix $I + T$ is primitive, hence for some k, $(I + T)^k > 0$; hence

$$z' (I + T)^k = \{\hat{x}' (I + T)^k\}\, T - r\, \{\hat{x}' (I + T)^k\} > \mathbf{0}$$

which contradicts the definition of r; (b) is then proved as in Theorem 1.6 following; and the rest follows as before, except for the last part in (c) #

We shall henceforth call r the Perron–Frobenius eigenvalue of an irreducible T, and its corresponding positive eigenvectors, the Perron–Frobenius eigenvectors.

The above theorem does not answer in detail questions about eigenvalues λ such that $\lambda \neq r$ but $|\lambda| = r$ in the cyclic case.

The following auxiliary result is more general than we shall require immediately, but is important in future contexts.

THEOREM 1.6. (The Subinvariance Theorem). Let T be a non-negative irreducible matrix, s a positive number, and $y \geqslant \mathbf{0}, \neq \mathbf{0}$, a vector satisfying

$$Ty \leqslant sy.$$

Then (a) $y > \mathbf{0}$; (b) $s \geqslant r$, where r is the Perron–Frobenius eigenvalue of T. Moreover, $s = r$, if and only if $Ty = ry$.

Proof. Suppose at least one element, say the ith, of y is zero. Then since $T^k y \leqslant s^k y$ it follows that

$$\sum_{j=1}^{n} t_{ij}^{(k)} y_j \leqslant s^k\, y_i.$$

Now, since T is irreducible, for this i and any j, there exists a k such that $t_{ij}^{(k)} > 0$; and since $y_j > 0$ for some j, it follows that

$$y_i > 0$$

which is a contradiction. Thus $y > \mathbf{0}$. Now, premultiplying the relation $Ty \leqslant sy$ by $\hat{x}'$, a positive left eigenvector of T corresponding to r,

$$s\,\hat{x}'\,y \geqslant \hat{x}'\,Ty = r\,\hat{x}'\,y$$

i.e.

$$s \geqslant r.$$

Now suppose $Ty \leqslant ry$ with strict inequality in at least one place; then the preceding argument, on account of the strict positivity of Ty and ry, yields $r < r$, which is impossible. The implication $s = r$ follows from $Ty = sy$ similarly. #

In the sequel, any subscripts which occur should be understood as reduced modulo d, to bring them into the range $[0, d - 1]$, if they do not already fall in the range.

THEOREM 1.7. For a cyclic matrix T with period $d > 1$, there are present precisely d distinct eigenvalues λ with $|\lambda| = r$, where r is the Perron–Frobenius eigenvalue of T. These eigenvalues are: $r \exp i2\pi k/d$, $k = 0, 1, \ldots, d - 1$ (i.e. the d roots of the equation $\lambda^d - r^d = 0$).

Proof. Consider an arbitrary one, say the ith, of the primitive matrices:

$$Q_{i,\ i+1}, Q_{i+1,\ i+2}, \ldots, Q_{i+d-1,\ i+d}$$

occurring as diagonal blocks in the dth power, T^d, of the canonical form T_c of T (recall that T_c has the same eigenvalues as T), and denote by $r(i)$ its Perron–Frobenius eigenvalue, and by $y(i)$ a corresponding positive right eigenvector, so that

$$Q_{i,\ i+1}\, Q_{i+1,\ i+2} \cdots Q_{i+d-1,\ i+d}\, y(i) = r(i)\, y(i).$$

Now premultiply this by $Q_{i-1,\ i}$:

$$Q_{i-1,\ i}\, Q_{i,\ i+1}\, Q_{i+1,\ i+2} \cdots Q_{i+d-2,\ i+d-1}\, Q_{i+d-1,\ i+d}\, y(i) = r(i)\, Q_{i-1,\ i}\, y(i),$$

and since $Q_{i+d-1,\ i+d} \equiv Q_{i-1,\ i}$, we have

$$Q_{i-1,\ i}\, Q_{i,\ i+1}\, Q_{i+1,\ i+2} \cdots Q_{i+d-2,\ i+d-1}\, (Q_{i-1,\ i}\, y(i)) = r(i)\, (Q_{i-1,\ i}\, y(i))$$

whence it follows from Theorem 1.6 that $r(i) \geqslant r(i-1)$. Thus

$$r(0) \geqslant r(d-1) \geqslant r(d-2) \ldots \geqslant r(0),$$

so that, for all i, $r(i)$ is constant, say $\tilde{r}$, and so there are precisely d dominant eigenvalues of T^d, all the other eigenvalues being strictly smaller in modulus. Hence, since the eigenvalues of T^d are dth powers of the eigenvalues of T, there must be precisely d dominant roots of T, and all must be dth roots of $\tilde{r}$. Now, from Theorem 1.5, the positive dth root is an eigenvalue of T and is r. Thus every root λ of T such that $|\lambda| = r$ must be of the form

$$\lambda = r \exp i(2\pi k/d),$$

where k is one of $0, 1, \ldots, d-1$, and there are d of them.

It remains to prove that there are no coincident eigenvalues, so that in fact all possibilities $r \exp i(2\pi k/d)$, $k = 0, 1, \ldots, d-1$ occur.

Suppose that y is a positive $(n \times 1)$ right eigenvector corresponding to the Perron–Frobenius eigenvalue r of T_c (i.e. T written out in canonical form), and let y_j, $j = 0, \ldots, d-1$ be the subvector of components corresponding to subclass C_j.

Thus
$$y' = [y'_0, y'_1, \ldots, y'_{d-1}]$$
and
$$Q_{j,\, j+1}\, y_{j+1} = r\, y_j.$$

Now, let $\bar{y}_k$, $k = 0, 1, \ldots, d-1$ be the $(n \times 1)$ vector obtained from y by making the transformation

$$y_j \to \exp \mathrm{i} \left(\frac{2\pi jk}{d}\right) y_j$$

of its components as defined above. It is easy to check that $\bar{y}_0 = y$, and indeed that $\bar{y}_k$, $k = 0, 1, \ldots, d-1$ is an eigenvector corresponding to an eigenvalue $r \exp \mathrm{i}\, (2\pi k/d)$, as required. This completes the proof of the theorem. #

We note in conclusion the following corollary on the structure of the eigenvalues, whose validity is now clear from the immediately preceding.

COROLLARY. If $\lambda \neq 0$ is any eigenvalue of T, then the numbers $\lambda \exp \mathrm{i}\, (2\pi k/d)$, $k = 0, 1, \ldots, d-1$ are eigenvalues also. (Thus, rotation of the complex plane about the origin through angles of $2\pi/d$ carries the set of eigenvalues into itself.)

Bibliography and discussion

1.1 and 1.4

The exposition of the chapter centres on the notion of a primitive non-negative matrix as the fundamental notion of non-negative matrix theory. The approach seems to have the advantage of proving the fundamental theorem of non-negative matrix theory at the outset, and of avoiding the slight awkwardness entailed in the usual definition of irreducibility merely from the permutable structure of T.

The fundamental results (Theorems 1.1, 1.5 and 1.7) are basically due to Perron (1907) and Frobenius (1908, 1909, 1912), Perron's contribution being associated with strictly positive T. Many modern expositions tend to follow the simple and elegant paper of Wielandt (1950) (whose approach was anticipated in part by Lappo-Danilevskii (1934)); see e.g. Cherubino (1957), Gantmacher (1959) and Varga (1962). This is essentially true also of our proof of Theorem 1.1 (= Theorem 1.5) with some slight simplifications, especially in the proof of part (*e*), under the influence of the well-known paper of Debreu & Herstein (1953), which deviates otherwise from Wielandt's treatment also in the proof of (*a*). (The proof of Corollary 1 of Theorem 1.1 also follows Debreu & Herstein.)

The proof of Theorem 1.7 is not, however, associated with Wielandt's approach, due to an attempt to bring out, again, the primacy of the primitivity property. The last part of the proof (that all dth roots of r are involved), as well as the corollary, follows Romanovsky (1936). The possibility of evolving §1.4 in the present manner depends heavily on §1.3.

For other approaches to the Perron–Frobenius theory see Bellman (1960, Chapter 16), Brauer (1957*b*), Fan (1958), Householder (1958), Karlin (1959, §8.2; 1966, Appendix), and Samelson (1957). Some of these references do not deal with the most general case of an irreducible matrix, containing restrictions of one sort or another.

1.2 and 1.3

The development of these sections is motivated by probabilistic considerations from the theory of Markov chains, where it occurs in connection with *stochastic* non-negative matrices $P = \{p_{ij}\}$, $i, j = 1, 2, \ldots$, with $p_{ij} \geqslant 0$ and

$$\sum_j p_{ij} = 1, \qquad i = 1, 2, \ldots.$$

In this setting the classification theory is essentially due to Kolmogorov (1936); an account may be found in the somewhat more general exposition of Chung (1967, Chapter 1, §3), which our exposition tends to follow.

Just as in the case of stochastic matrices, the corresponding exposition *is not restricted to finite matrices* (this in fact being the reason for the development of this kind of classification in the probabilistic setting), and virtually all of the present exposition goes through for infinite non-negative matrices T, so long as all powers T^k, $k = 1, 2, \ldots$ exist (with an obvious extension of the rule of matrix multiplication of finite matrices). This point is taken up again to a limited extent in Chapters 5 and 6, where infinite T are studied.

The reader acquainted with graph theory will recognize its relationship with the notion of path diagrams used in our exposition. For development along the lines of graph theory see Rosenblatt (1957), the brief account in Varga (1962, Chapters 1 and 2), Paz (1963) and Gordon (1965, Chapter 1). The relevant notions and usage in the setting of non-negative matrices implicitly go back at least to Romanovsky (1936).

Another development, not explicitly graph theoretical, is given in the papers of Pták (1958) and Pták & Sedlaček (1958); and it utilized to some extent in §2.4 of the next chapter.

Finite stochastic matrices and finite Markov chains will be briefly treated in Chapter 4. The general infinite case will be taken up in Chapter 5.

Exercises

1.1. Find all essential and inessential classes of a non-negative matrix with incidence matrix:

$$T = \begin{bmatrix} 0 & 0 & 0 & 0 & 0 & 1 & 0 & 0 \\ 0 & 0 & 0 & 1 & 1 & 0 & 0 & 0 \\ 1 & 0 & 0 & 0 & 0 & 0 & 0 & 1 \\ 0 & 0 & 0 & 0 & 0 & 0 & 0 & 1 \\ 0 & 1 & 0 & 0 & 1 & 0 & 0 & 0 \\ 0 & 0 & 1 & 1 & 0 & 0 & 0 & 0 \\ 0 & 0 & 0 & 0 & 1 & 0 & 1 & 1 \\ 0 & 0 & 0 & 0 & 0 & 1 & 0 & 0 \end{bmatrix}$$

Find the periods of all self-communicating classes, and write the matrix T in full canonical form, so that the matrices corresponding to all self-communicating classes are also in canonical form.

1.2. Keeping in mind Lemma 1.1, construct a non-negative matrix T whose index set contains no essential class, but has, nevertheless, a self-communicating class.

1.3. Given $T = \{t_{i,j}\}$, $i, j = 1, 2, \ldots, n$ is a non-negative irreducible matrix, show that for any fixed i and j, there exists a sequence $k_1, k_2, \ldots, k_r$ such that

$$t_{i,k_1} t_{k_1,k_2} \ldots t_{k_{r-1},k_r} t_{k_r,j} > 0$$

where $r \leqslant n-1$. (Consider separately the cases $i = j$ and $i \neq j$.)

If in addition it is given that $t_{i,i} > 0$, $i = 1, \ldots, n$ show that $T^{n-1} > 0$.

(Wielandt, 1950; Herstein, 1954)

1.4. Given $T = \{t_{i,j}\}$, $i, j = 1, 2, \ldots, n$ is a non-negative matrix, suppose that for some power $m \geqslant 1$, $T^m = \{t_{i,j}^{(m)}\}$ is such that

$$t_{i,i+1}^{(m)} > 0,\ i = 1, 2, \ldots, n-1, \text{ and } t_{n,1}^{(m)} > 0.$$

Show that: T is irreducible; and (by example) that it may be periodic.

1.5. By considering the vector $x' = (\alpha, \alpha, 1-2\alpha)$, suitably normed, when: (i) $\alpha = 0$, (ii) $0 < \alpha < \frac{1}{2}$, and the matrix

$$\begin{bmatrix} 0 & 1 & 0 \\ 0 & 0 & 2 \\ 1 & 0 & 2 \end{bmatrix},$$

show that $r(x)$, as defined in the proof of Theorem 1.1 is not continuous in $x \geqslant 0$, $x'x = 1$.

(Schneider, 1958)

1.6.† If r is the Perron–Frobenius eigenvalue of an irreducible matrix $T = \{t_{ij}\}$, show that for any vector $x \in \mathscr{P}$, where $\mathscr{P} = \{x; x > 0\}$

$$\min_i \frac{\sum_j t_{ij} x_j}{x_i} \leqslant r \leqslant \max_i \frac{\sum_j t_{ij} x_j}{x_i}.$$

(Collatz, 1942)

† Exercises 1.6 to 1.8 have a common theme.

1.7. Show, in the situation of Exercise 1.6, that equality on either side implies equality on both; and by considering when this can happen show that r is the supremum of the left hand side, and the infimum of the right hand side, over $x \in \mathscr{P}$, and is actually attained as both supremum and infimum for vectors in $\mathscr{P}$.

1.8. In the framework of Exercise 1.6, show that

$$\max_{x \in \mathscr{P}} \left\{ \min_{y \in \mathscr{P}} \frac{y' T x}{y' x} \right\} = r = \min_{y \in \mathscr{P}} \left\{ \max_{x \in \mathscr{P}} \frac{y' T x}{y' x} \right\}.$$

(Birkhoff and Varga, 1958)

1.9.† Let B be a matrix with possibly complex elements and denote by B_+ the matrix of moduli of elements of B and β an eigenvalue of B. Let T be irreducible and such that $0 \leqslant B_+ \leqslant T$. Show that $|\beta| \leqslant r$; and moreover that $|\beta| = r$ implies $B_+ = T$, where r is the Perron–Frobenius eigenvalue of T.

(Frobenius, 1909)

1.10. If, in Exercise 1.9, $|\beta| = r$, so that $\beta = r\, e^{i\theta}$, say, it can be shown (Wielandt, 1950) that B has the representation

$$B = e^{i\theta} D T D^{-1}$$

where D is a diagonal matrix whose diagonal elements have modulus one. Shown as consequences:

(i) that if $|\beta| = r$, $B_+ = T$;

(ii) that given there are d dominant eigenvalues of modulus r for a given periodic irreducible matrix of period d, they must in fact all be simple, and take on the values $r \exp i(2\pi j/d)$, $j = 0, 1, \ldots, d-1$. (Put $B = T$ in the representation.)

1.11. Let T be an irreducible non-negative matrix and E a non-zero non-negative matrix of the same size. If x is a positive number, show that $A = x E + T$ is irreducible, and that its Perron–Frobenius eigenvalue may be made to equal any positive number exceeding the Perron–Frobenius eigenvalue r of T by suitable choice of x.

(Consider first, for orientation, the situation where at least one diagonal element of E is positive. Make eventual use of the continuity of the eigenvalues of A with x.) (Birkhoff & Varga, 1958)

1.12. If $T \geqslant 0$ is any square non-negative matrix, use the canonical form of T to show that the following weak analogue of the Perron–Frobenius Theorem holds: there exists an eigenvalue ρ such that

(a') ρ real, $\geqslant 0$;
(b') with ρ can be associated non-negative left and right eigenvectors;
(c') $\rho \geqslant |\lambda|$ for any eigenvalue λ of T;
(e') if $0 \leqslant B \leqslant T$ and β is an eigenvalue of B, then $|\beta| \leqslant \rho$.

† Exercises 1.9 to 1.11 have a common theme.

(In such problems it is often useful to consider a sequence of matrices each $\geqslant T$ and converging to T elementwise $\{$particularly in relation to (b') here$\}$ – Debreu and Herstein (1953).)

1.13. Show in relation to Exercise 1.12, that $\rho > 0$ if and only if T contains a *cycle* of elements. (Ullman, 1952)

1.14. Use the relevant part of Theorem 1.4, in conjunction with Theorem 1.2, to show that for an irreducible T with Perron–Frobenius eigenvalue r, as $k \to \infty$

$$s^{-k}\, T^k \to 0$$

if and only if $s > r$; and if $0 < s < r$, for each pair (i, j)

$$\limsup_{k \to \infty} s^{-k}\, t_{ij}^{(k)} = \infty.$$

Hence deduce that the power series

$$T_{ij}(z) = \sum_{k=0}^{\infty} t_{ij}^{(k)} z^k$$

have common convergence radius $R = r^{-1}$ for each pair (i, j). (This result is relevant to the development of the theory of countable irreducible T in Chapter 6.)

2 Some secondary theory with emphasis on irreducible matrices, and applications

In this chapter we survey briefly some of the theory which has arisen out of deeper investigation, and generalization, of various aspects of the Perron–Frobenius structure of a non-negative *irreducible* matrix T, with Perron–Frobenius eigenvalue r. Some of the material is of particular relevance in certain application (e.g. mathematical economics, numerical analysis, demography); these are also briefly discussed.

It is possible to extend several of the results to a reducible matrix T, via canonical form; some such discussion is deferred to the exercises.

2.1 The equations: $(sI - T)\,\boldsymbol{x} = \boldsymbol{c}$

In a well-known mathematical–economic setting, to be discussed shortly, it is desired to investigate conditions ensuring positivity ($\boldsymbol{x} > \boldsymbol{0}$) of solutions to the equation system

$$(sI - T)\,\boldsymbol{x} = \boldsymbol{c}$$

for any $\boldsymbol{c} \geqslant \boldsymbol{0}, \neq \boldsymbol{0}$. Closely related to this is the question: for what values of s do we have $(sI - T)^{-1} > 0$?

THEOREM 2.1 A necessary and sufficient condition for a solution $\boldsymbol{x}$ ($\boldsymbol{x} \geqslant \boldsymbol{0}, \neq \boldsymbol{0}$) to the equations

$$(sI - T)\,\boldsymbol{x} = \boldsymbol{c} \tag{2.1}$$

to exist for any $\boldsymbol{c} \geqslant \boldsymbol{0}, \neq \boldsymbol{0}$ is that $s > r$. In this case there is only one solution $\boldsymbol{x}$, which is strictly positive and given by

$$\boldsymbol{x} = (sI - T)^{-1}\,\boldsymbol{c}.$$

Proof. Suppose first that for some $\boldsymbol{c} \geqslant \mathbf{0}, \neq \mathbf{0}$ a non-negative non-zero solution to (2.1) exists. Then

$$c + Tx = sx$$

i.e.

$$Tx \leqslant sx$$

with strict inequality for at least one element. This is impossible for $s \leqslant 0$, and if $s > 0$, Theorem 1.6 implies $s > r$.

Now suppose $s > r$. Then, since $T^k/s^k \to 0$ as $k \to \infty$ (see Exercise 1.14), it follows that

$$(sI - T)^{-1} = s^{-1}(I - s^{-1}T)^{-1} = s^{-1} \sum_{k=0}^{\infty} (s^{-1}T)^k$$

exists, from Lemma B.1 of Appendix B; and moreover, since for any pair i, j, $t_{ij}^{(k)} > 0$ for some $k = k(i,j)$ by irreducibility, it follows that the right hand side is a strictly positive matrix. Hence

$$(sI - T)^{-1} > 0$$

so that

$$(sI - T)^{-1} \boldsymbol{c} > \mathbf{0} \text{ for any } \boldsymbol{c} \geqslant \mathbf{0}. \neq \mathbf{0},$$

and clearly

$$\boldsymbol{x} = (sI - T)^{-1} \boldsymbol{c}. \quad \#$$

COROLLARY 1. Of those real numbers s for which it exists, $(sI - T)^{-1} > 0$ if and only if $s > r$.

COROLLARY 2. If $s = 1$, then the necessary and sufficient condition stated for the theorem becomes $r < 1$.

COROLLARY 3. If $s = 1$, then $r < 1$ if none of the row (column) sums of T exceed unity, and at least one is less than unity.

Proof. Follows directly from Corollary 1 of the Perron–Frobenius Theorem (Theorem 1.1). #

THEOREM 2.2. A condition equivalent to $s > r$ is that

$$\Delta_i(s) > 0, \quad i = 1, 2, \ldots, n \tag{2.2}$$

where $\Delta_i(s)$ is the principal minor of $(sI - T)$ which consists of the first i rows and columns of $(sI - T)$.

Proof.† Assume first that $s > r$. Then

$$\Delta_n = \det(sI - T) = \phi(s),$$

in our previous notation, exceeds zero since it is known that $\phi(x) \to \infty$ as $x \to \infty$, and s lies beyond the largest real root r of T. Moreover, since s must

† We shall write in the proof Δ_i rather than $\Delta_i(s)$ for simplicity, since s is fixed.

exceed the maximal modulus real non-negative eigenvalue of the matrix formed by the first i rows and columns of T, $i < n$ (see Exercise 1.12), it must similarly follow that $\Delta_i > 0$ for $i = 1, 2, \ldots, n-1$, if $n > 1$.

Assume now that (2.2) holds for some fixed real s. Since each of the Δ_i is a continuous function of the entries of T, it follows that it is possible to replace all the zero entries of T by sufficiently small positive entries to produce a positive matrix $\bar{T}$ with Perron–Frobenius eigenvalue $\bar{r} > r$ by Theorem 1.1(e), for which still $\bar{\Delta}_i > 0$, $i = 1, 2, \ldots, n$. Thus if we can prove that $s > \bar{r}$, this will suffice for what is required.

It follows, then, that it suffices to prove that

$$\Delta_i > 0,\ i = 1, 2, \ldots, n \text{ implies } s > r$$

for *positive* matrices T, which is what we now assume about T. We proceed by induction on the dimension n of the matrix T. If $n = 1$, $\Delta_i > 0$ implies $s > t_{11} \equiv T \equiv r$.

Suppose now the proposition (2.2) is true for matrices of dimension n; and for a matrix T of dimension $(n + 1)$ assume

$$\Delta_i > 0,\ i = 1, 2, \ldots, n + 1.$$

If r_n is the Perron–Frobenius eigenvalue of the $(n \times n)$ positive matrix ${}_{(n)}T$ which arises out of crossing out the last row and column of T, we have by induction that $s > r_n$. Let

$$k_{n+1} \equiv \Delta_n / \Delta_{n+1} > 0,$$

and consider the unique solution $\boldsymbol{x} = \boldsymbol{\alpha}_{n+1}$ of the system

$$(sI - T)\boldsymbol{x} = f_{n+1} \tag{2.3}$$

where $f_{n+1} = (0, 0, \ldots, 0, 1)$. In the first instance, since

$$\alpha_{n+1} = (sI - T)^{-1} f_{n+1},$$

it must follow that the $n + 1$ element of $\boldsymbol{\alpha}_{n+1}$ is k_{n+1}, since this is the $(n + 1, n + 1)$ element of $(sI - T)^{-1}$. If we rewrite

$$\boldsymbol{\alpha}_{n+1} = (\boldsymbol{\alpha}_n', k_{n+1})',$$

it follows that $\boldsymbol{x} = \boldsymbol{\alpha}_n$ must satisfy

$$(s\,{}_{(n)}I - {}_{(n)}T)\,\boldsymbol{x} = \boldsymbol{\sigma}$$

where $$\sigma_n = t_{i,\,n+1} k_{n+1} > 0$$

from the first n equations of (2.3). But, since $s > r_n$, Theorem 2.1 implies that the unique solution, viz. $\boldsymbol{\alpha}_n$, is strictly positive. Hence $\boldsymbol{\alpha}_{n+1} > \boldsymbol{0}$ also, and the Subinvariance Theorem applied to (2.3) now implies $s > r_{n+1}$, as required. #

THEOREM 2.3 If $s \geqslant \max_i \sum_{j=1}^{n} t_{ij}$, and $c_{ij}(s)$ is the cofactor of the ith row and jth column of $sI - T$, then $c_{ii}(s) \geqslant c_{ij}(s) > 0$, all i, j.

Proof.† Adj $(sI - T)$ is the transposed matrix of cofactors c_{ij}, and is certainly positive if $s > r$ by Theorem 2.1 above, since

$$0 < (sI - T)^{-1} = \text{Adj}\,(sI - T)/\phi(s)$$

where $\phi(s) > 0$. Further, if $s = r$ it is also positive, by Corollary 2 to the Perron–Frobenius Theorem. Now, since Corollary 1 of the Perron–Frobenius Theorem asserts that $r \leqslant \max_i \sum_j t_{ij}, \leqslant s$ by assumption, it follows that all the cofactors are positive.

(i) Consider first the case $s > \max_i \sum_j t_{ij}$.

Now, replace any zero entries of T by a small positive number δ to form a new positive matrix $\bar{T} = \{\bar{t}_{ij}\}$ but such that still $s > \max_i \sum_j \bar{t}_{ij}$. If we can prove $\bar{c}_{ii} \geqslant \bar{c}_{ij}$ all i, j in this case, this suffices, for, by continuity in δ, letting $\delta \to 0+$, $c_{ij} \geqslant c_{ii}$.

Thus it suffices to consider a totally positive matrix $T = \{t_{ij}\}$ which we shall now do. Take i, and j, $j \neq i$ fixed (but arbitrary). Replace all elements of the ith row of T : by zeroes, except (i, i)th and (i, j)th where we put $s/2$. Call the new matrix U; it is clearly irreducible, and moreover has all row sums not exceeding s, and all but the ith less than s; thus its Perron–Frobenius eigenvalue $r_U < s$ by Corollary 1 to the Perron–Frobenius Theorem and so

$$0 < \det\,(sI - U) = -\frac{s}{2}c_{ij} + \frac{s}{2}c_{ii}$$

expanding the determinant by the ith row of $sI - U$ and recalling that the cofactors remain the same as for $sI - T$.

Therefore $$c_{ij} < c_{ii}$$

which is as required.

(ii) $s = \max_i \sum_j t_{ij}$. Take $\delta > 0$ and consider $s + \delta$ in place of s in $(sI - T)$. Then since $c_{ij}(s + \delta) \leqslant c_{ii}(s + \delta)$ for all $\delta > 0$, from part (i), it follows by continuity that, letting $\delta \to 0$, $c_{ij} \leqslant c_{ii}$ as required. #

COROLLARY 1. If in fact $T > 0$ then the conclusion can be strengthened to $c_{ij}(s) < c_{ii}(s)$ all i, j.

COROLLARY 2. If $s \geqslant \max_j \sum_{i=1}^{n} t_{ij}$ then $c_{ji}(s) \leqslant c_{ii}(s)$ for all i, j.

Proof. T', the transpose of T, satisfies the conditions of Theorem 2.3. #

The open Leontief model

Consider an economy in which there are n industries, each industry producing exactly one kind of good (commodity). Let x_i be the output of the ith

† We shall write in the proof c_{ij} rather than $c_{ij}(s)$, since s is fixed.

industry, and $t_{ij} \geqslant 0$ the input of commodity i per unit output of commodity j. It follows then that $t_{ij}\,x_j$ is the amount of the output of commodity i absorbed in the production of commodity j, and the excess

$$c_i = x_i - \sum_{j=1}^{n} t_{ij}\,x_j,\; i = 1, \ldots, n$$

is the amount of commodity i available for outside use. Thus the vector $x = \{x_i\}$ may be interpreted as a 'supply' vector, and the vector $c = \{c_i\}$ as a 'demand' vector. A question of importance, therefore, is that of when for a given $c \geqslant 0, \neq 0$, is there a solution $x \geqslant 0, \neq 0$ to the system

$$(I - T)\,x = c. \tag{2.4}$$

Theorems 2.1 and 2.2 above answer this question in the special case $s = 1$, under the assumption of irreducibility of T. Moreover when the theorems hold, the positivity of $(I - T)^{-1}$, and the fact that $x = (I - T)^{-1}\,c$, guarantees that an increase in demand of even one good, increases the output of all goods.

If the necessary and sufficient condition $r < 1$ is replaced by the stronger assumption that for all i,

$$\sum_{j=1}^{n} t_{ij} \leqslant 1, \tag{2.5}$$

with strict inequality for at least one i, then Theorem 2.3 gives the additional result that if only the demand for commodity j increases then the output of commodity j increases by the greater percentage, though all outputs increase.†

Whereas it is difficult to ascribe a direct meaning in economic terms to the condition (2.5) within the context of the present model (in contrast to certain other contexts), the dual condition that for all j

$$\sum_{i=1}^{n} t_{ij} \leqslant 1,$$

with strict inequality for at least one j, may be given a meaningful interpretation: if commodities are measured in 'dollars worth' units, this is the condition that no industry operates at a loss, and at least one operates at a profit, in terms of internal ('non-labour') factors.

Example: Consider the matrix

$$T_1 = \begin{bmatrix} \frac{2}{11} & \frac{6}{11} \\ \frac{8}{11} & \frac{4}{11} \end{bmatrix}.$$

T_1 is positive, hence irreducible. The top row sum is less than unity, the bottom one exceeds unity, so that the condition of Theorem 2.3 is not satisfied.

† See Exercise 2.2.

However the column sums are both $\frac{10}{11} < 1$ so that Corollary 2 of Theorem 1.3 does apply for $s = 1$. In fact $r = \frac{10}{11}$ by Corollary 1 of the Perron–Frobenius Theorem, so that Theorem 2.1 holds with $s = 1$. In fact

$$I - T_1 = \begin{bmatrix} \frac{9}{11} & \frac{-6}{11} \\ \frac{-8}{11} & \frac{7}{11} \end{bmatrix}, \quad (I - T_1)^{-1} = \frac{11}{15}\begin{bmatrix} 7 & 6 \\ 8 & 9 \end{bmatrix}$$

which agrees with the assertions of these results.

Note, also, that *if the demand for commodity 1 only increases,* by a single unit, the supply vector increases by $\left[\frac{77}{15}, \frac{88}{15}\right]$, *which is a greater increase in supply of commodity 2 than of commodity 1.*

On the other hand if the matrix

$$T_2 = T_1' = \begin{bmatrix} \frac{2}{11} & \frac{8}{11} \\ \frac{6}{11} & \frac{4}{11} \end{bmatrix}$$

is considered, then all of Theorems 2.1 to 2.3 hold and

$$(I - T_2)^{-1} = [(I - T_1)^{-1}]' = \frac{11}{15}\begin{bmatrix} 7 & 8 \\ 6 & 9 \end{bmatrix}$$

so that unit increase in demand of either commodity, the other being held constant, forces a greater increase in supply of that commodity than the other, as expected.

We thus pause to note that (i) it is not possible to increase the demand vector $\boldsymbol{c}$ even in one component while keeping any element of the supply vector $\boldsymbol{x}$ fixed; (ii) it is not necessarily true that an increase in demand for a single commodity forces a greater (absolute or proportional) increase in the supply of that commodity compared to others.†

The economic model just discussed is called the open Leontief model; other economic interpretations may also be given to the same mathematical framework.

We shall mention a *dynamic* model, whose *static* version is formally (i.e. mathematically) identical with the Leontief set-up. Let the elements of $\boldsymbol{x}_k$

† cf. Morishima (1964), Chapter 1, Theorems 6 and 8.

denote the output at time stage k of various industries as before; let β, $0 < \beta \leqslant 1$ be the proportion (the same for each industry) of output at any stage available for internal use, a proportion $1-\beta$ being needed to meet external demand; let t_{ij} be the amount of output of industry i per unit of input of industry j, at the next time stage, and let $\boldsymbol{c} \geqslant \boldsymbol{0}, \neq \boldsymbol{0}$, be an 'external input' vector, the same at all time stages. Then

$$\boldsymbol{x}_{k+1} = \beta T \boldsymbol{x}_k + \boldsymbol{c}.$$

The general solution of this difference equation is

$$\boldsymbol{x}_k = (\beta T)^k \boldsymbol{x}_0 + \left\{ \sum_{i=1}^{k-1} (\beta T)^i \right\} \boldsymbol{c}.$$

If T is assumed irreducible, as usual, then if $(\beta T)^i \to 0$ as $i \to \infty$, $\boldsymbol{x}_k$ converges elementwise to the solution

$$\boldsymbol{x} = (I - \beta T)^{-1} \boldsymbol{c}$$

(see Lemma B.1) of the 'stationary system'. Necessary and sufficient for $(\beta T)^i \to 0$ is $r < \beta^{-1}$ (see Exercise 1.14) which is familiar from Theorems 2.1 and 2.2 with $s = \beta^{-1}$.

Bibliography and discussion to §2.1

Theorem 2.1, in the form of its Corollary 1, goes back to Frobenius (1912) for positive T.

The form of necessary and sufficient condition embodied in Theorem 2.2 is implicit in the paper of Hawkins & Simon (1949), who deal with positive T; and is explicit in a statement of Georgescu-Roegen (1951). The condition is thus often called the Hawkins–Simon condition in mathematico-economic contexts. Theorem 2.2 in fact holds for a non-negative T which is not necessarily irreducible† as was demonstrated by Gantmacher (1959, pp. 85–9; original Russian version: 1954) who himself attributes this result to Kotelyanskii (1952) whose proof he imitates, although in actual fact Kotelyanskii (*a*) considered $T > 0$, (*b*) proved a (non-trivial) variant‡ of Theorem 2.2 where every sign '$>$' in its statement is replaced by '$\geqslant$'. The Kotelyanskii–Gantmacher assertion was also obtained by Burger (1957) who followed the method of the Hawkins–Simon paper; and by Fan (1958). Further, Morishima (1964, Chapter 1) gives a proof of Theorem 2.2 which, however, makes the implicit assumption that if the last row and column of an irreducible T are deleted, then the matrix remaining is also irreducible. The virtue of the Hawkins–Simon condition, as well as the sufficient condition given in Corollary 3 of Theorem 2.1, is that such conditions may be checked fairly easily from the form of the matrix T, without the necessity of calculating r.

† See Exercise 2.4.

‡ See Exercise 2.12 (in §2.3).

Finally, in relation to Theorem 2.2, it is relevant to mention that less powerful assertions of similar form, but involving the non-negativity of *all* (not just leading) principal minors, begin with Frobenius (1908).

Theorem 2.3 is generally attributed to Metzler (1945, 1951). Debreu & Herstein (1953) give a proof of its Corollary 1 which the proof of the present theorem imitates. (The present statement appears to be marginally more general than those usually encountered.)

A simple direct discussion of linear models of production in econometrics, including the Leontief model, is given by Gale (1960, Chapter 9). The reader interested in an emphasis on non-negative matrix theory in connection with the Leontief model, and alternative interpretations of it, should consult Solow (1952) and Karlin (1959, §§8.1–8.3). We also mention Kemeny & Snell's (1960, §7.7) discussion of this model, in which $\sum_i t_{ij} \leqslant 1$ all j, but T is not necessarily irreducible, in a Markov chain framework closely related to our development in Chapter 1.

Finally, for a very detailed and extensive mathematical discussion pertaining to the properties of the matrices $sI - T$, and further references, the reader should consult the articles of Wong and of Woodbury in Morgenstern (1954).

Exercises on §2.1

In Exercises 2.1 to 2.3, which have a common theme, T is an irreducible non-negative matrix with Perron–Frobenius eigenvalue r.

2.1. Let $s > 0$. Show that a necessary and sufficient condition for $r < s$ is the existence of a vector $\boldsymbol{x}$, $\boldsymbol{x} \geqslant \boldsymbol{0}$, $\neq \boldsymbol{0}$ such that:

$$T\boldsymbol{x} \leqslant s\boldsymbol{x}$$

with strict inequality in at least one position.
[The condition $T\boldsymbol{x} \leqslant \boldsymbol{x}$, $\boldsymbol{x} \geqslant \boldsymbol{0}$, $\neq \boldsymbol{0}$ is sometimes imposed as a fundamental assumption in simple direct discussions of linear models; in the Leontief model it is tantamount to an assertion that there is at least one demand $\boldsymbol{c}$, $\boldsymbol{c} \geqslant \boldsymbol{0}$, $\neq \boldsymbol{0}$ which can be met, i.e. the system is 'productive'.]

2.2. Show that if

$$s \geqslant \sum_{j=1}^{n} t_{ij},$$

with strict inequality for at least one i, then Theorem 2.1 holds, and, moreover, if in the equation system (2.1) $\boldsymbol{c} \geqslant \boldsymbol{0}$, $\neq \boldsymbol{0}$ is increased in the jth component only, then the greatest percentage increase in $\boldsymbol{x}$ is in the jth component also.

2.3. Suppose T is any square non-negative matrix. Show that a *necessary* condition for $(sI - T)^{-1} > 0$ for some s is that T be irreducible, i.e. the situation $(sI - T)^{-1} > 0$ *may only occur if T is irreducible.* (Suppose initially that T has the partitioned form

$$T = \begin{bmatrix} A & 0 \\ C & B \end{bmatrix}$$

where A and B are square and irreducible, 0 is as usual a zero matrix, and $(sI - T)^{-1}$ exists.)

2.4.† Suppose T is any square non-negative matrix, with dominant eigenvalue ρ (guaranteed by Exercise 1.12). Show that $\Delta_i > 0, i = 1, \ldots, n$ if and only if $s > \rho$. (Kotelyanskii, 1952)
[Since it is easily shown that $(sI - T)^{-1} \geqslant 0$ if $s > \rho$ – see e.g. Debreu & Herstein (1953) – it follows that the Hawkins–Simon condition ensures *non-negativity* of solution to the Leontief system (2.1), taking $s = 1$, (i.e. (2.4)), even if T is reducible, for any 'demand' vector $\boldsymbol{c} \geqslant \boldsymbol{0}, \neq \boldsymbol{0}$.]

2.5. Suppose, in the setting of Exercise 2.4, assuming all cofactors $c_{ij}(1) \geqslant 0$,

$$1 \geqslant \max_i \sum_{j=1}^{n} t_{ij}$$

and $\rho < 1$, use induction to show

$$1 > \Delta_1 (1) > \Delta_2 (1) > \ldots > \Delta_n (1) > 0$$

(Wong, 1954)

2.6. Suppose, in the context of the open Leontief model, that each industry supplies just one industry which differs from itself; and that each of the industries has only one supplier. Show by example that T is not necessarily irreducible. Is it possible to have T primitive in this context? In general, can any of the indices of T be inessential?

2.2 Iterative methods for solution of certain linear equation systems

We consider in this section a system of equations

$$A\boldsymbol{x} = \boldsymbol{b} \tag{2.6}$$

where $A = \{a_{ij}\}, i, j = 1, \ldots, n$ has all its *diagonal elements positive,* and $\boldsymbol{b}$ is some given column vector. The matrix A may be expressed in the form

$$A = D - E - F$$

where $D = \text{diag}\{a_{11}, a_{22}, \ldots, a_{nn}\}$ and E and F are respectively strictly lower and upper triangular $n \times n$ matrices, whose entries are the negatives of the entries of A respectively below and above the main diagonal of A. The equation system may, thus, be rewritten,

$$\{I - D^{-1} (E + F)\} \boldsymbol{x} = D^{-1} \boldsymbol{b}.$$

† Exercises 2.4 and 2.5 have a common theme.

We shall make the assumption that

$$T_1 = D^{-1}(E + F)$$

is non-negative, and write for convenience

$$L = D^{-1}E,\ U = D^{-1}F \text{ and } \boldsymbol{k} = D^{-1}\boldsymbol{b}.$$

Thus we are concerned basically with the equation system

$$(I - T)\boldsymbol{x} = \boldsymbol{k} \tag{2.7}$$

where T is non-negative. Thus the equation system considered in §2.1 falls into this framework, as do any systems where A has

(i) positive diagonal and non-positive off-diagonal entries; or

(ii) negative diagonal and non-negative off-diagonal entries (the case (ii) occurs in the theory of continuous parameter Markov processes).

There are two well-known iterative methods for attempting to find a solution $\boldsymbol{x}$ (if one exists) of (2.7) corresponding respectively to the two ways

$$\boldsymbol{x} = (L + U)\boldsymbol{x} + \boldsymbol{k}$$

$$(I - L)\boldsymbol{x} = U\boldsymbol{x} + \boldsymbol{k}$$

of rewriting the equation system (2.7).

Jacobi iteration: The $(m + 1)$th approximation $\boldsymbol{x}(m + 1)$ is derived from the mth by

$$\boldsymbol{x}(m + 1) = (L + U)\boldsymbol{x}(m) + \boldsymbol{k}.$$

$T_1 = (L + U)$ is *the Jacobi matrix* of the system.

Gauss–Seidel iteration: $\boldsymbol{x}(m + 1)$ is related to $\boldsymbol{x}(m)$ by

$$(I - L)\boldsymbol{x}(m + 1) = U\boldsymbol{x}(m) + \boldsymbol{k}$$

i.e. $$\boldsymbol{x}(m + 1) = (I - L)^{-1}U\boldsymbol{x}(m) + (I - L)^{-1}\boldsymbol{k}.$$

$(I - L)^{-1}$ exists, since L is strictly lower triangular: in fact clearly, because of this last fact,

$$L^k = 0,\ k \geqslant n, \text{ so that}$$

$$(I - L)^{-1} = \sum_{k=0}^{n-1} L^k$$

(see Lemma B.1 of Appendix B).

The matrix $T_2 = (I - L)^{-1}U$ *is the Gauss–Seidel matrix* of the system.

We shall assume that T_1 is irreducible, with Perron–Frobenius eigenvalue r_1; the matrix $T_2 = (I - L)^{-1}U$ is reducible,† so let r_2 denote its dominant (non-negative) eigenvalue in accordance with the content of Exercise 1.12. The significance of the following theorem concerning relative size of r_1 and

† See Exercise 2.7.

r_2 will be explained shortly. (Note that irreducibility of T_1 implies $L \neq 0, U \neq 0$.)

THEOREM 2.4. One and only one of the following relations is valid:

$$(1)\ 1 = r_2 = r_1,$$
$$(2)\ 1 < r_1 < r_2,$$
$$(3)\ 0 < r_2 < r_1 < 1.$$

Proof. Let x be a positive eigenvector corresponding to the Perron–Frobenius eigenvalue r_1 of T_1. Then

$$(L + U)\,x = r_1\,x$$

so that
$$(I - r_1^{-1}\,L)^{-1}\,(L + U)\,x = r_1\,(I - r_1^{-1}\,L)^{-1}\,x,$$

i.e.
$$(I - r_1^{-1}\,L)^{-1}\,U\,x = r_1\,x.$$

Thus r_1 is an eigenvalue and x an eigenvector of $(I - r_1^{-1}\,L)^{-1}\,U$ which is

$$\sum_{k=0}^{n-1} (r_1^{-1}\,L)^k\,U.$$

We deduce immediately (since the presence of the r_1 does not change the position of the positive or zero elements) that r_2, the dominant eigenvalue of

$$(I - L)^{-1}\,U = \sum_{k=0}^{n-1} L^k\,U, \textit{ satisfies } r_2 > 0, \text{ since } r_1 > 0.$$

Further, if $r_1 \geq 1$, $(I - r_1^{-1}\,L)^{-1}\,U \leq (I - L)^{-1}\,U$;
it follows that $r_2 \geq r_1$, from the content of Exercise 1.12. In fact, since both incidence matrices are the same, a consideration of Theorem 1.1(*e*) reveals $r_1 > 1 \Rightarrow r_2 > r_1$.

Now, suppose that y, $y \geq \mathbf{0}, \neq \mathbf{0}$ is a right eigenvector corresponding to the eigenvalue r_2 of T_2. Thus

$$(I - L)^{-1}\,Uy = r_2\,y$$

so that
$$Uy = r_2\,(I - L)\,y,$$

i.e.
$$(r_2 L + U)y = r_2 y.$$

Since $r_2\,L + U$ must be irreducible (since $L + U$ is), r_2 must be its Perron–Frobenius eigenvalue, and $y > \mathbf{0}$ (see Theorem 1.6). If $r_2 > 1$, then

$$r_2\,L + U \leq r_2\,(L + U)$$

with strict inequality in at least one position, so that, from the Perron–Frobenius theory, $r_2 < r_2\,r_1$, i.e. $1 < r_1$; but if $r_2 = 1$, then $r_1 = 1$. On the other hand, if $r_2 < 1$

$$r_2\,L + U \leq L + U$$

with strict inequality in at least one position, so that $r_2 < r_1$.

Let us now summarize our deductions

(i) $r_1 \geqslant 1 \Rightarrow r_1 \leqslant r_2$; in fact $r_1 > 1 \Rightarrow r_1 < r_2$.
(ii) $r_2 > 1 \Rightarrow 1 < r_1$.
(iii) $r_2 = 1 \Rightarrow 1 = r_1$.
(iv) $r_2 < 1 \Rightarrow r_2 < r_1$.

The conclusion of the theorem follows. #

Returning now to the Jacobi and Gauss–Seidel methods, we may further investigate both methods by looking at the system, for fixed i:

$$x(m+1) = T_i x(m) + \delta_i, \quad i = 1, 2$$

where δ_i is some fixed vector. If a solution x to the system (2.7) exists then clearly the error vector at time $m + 1$, $\epsilon(m + 1)$ is given by:

$$\epsilon(m+1) \equiv x(m+1) - x = T_i\,[x(m) - x]$$

so that
$$\epsilon(m) = T_i^m\,[x(0) - x] = T_i^m\,\epsilon(0).$$

Thus if $T_i^m \to 0$ as $m \to \infty$ the ith iterative method results in *convergence* to the solution, which is then unique, in fact, since $T_i^m \to 0$ ensures that $(I - T_i)^{-1}$ exists. On the other hand, if $T_i^m \nrightarrow 0$, *even if* a unique solution x is known to occur, the ith iterative method will not in general converge to it.

Before passing on to further discussion, the reader may wish to note that the problem of convergence here in the case $i = 1$ is identical with that of convergence of the dynamic economic model discussed at the conclusion of §2.1. We have already seen there that since T_1 is irreducible, a necessary and sufficient condition for $T_i^m \to 0$ is $r_1 < 1$.

T_2 on the other hand is reducible, and, while we have not developed appropriate theory to cope with this problem, the Jordan canonical form of a matrix being usually involved, nevertheless the relevant result† yields an analogous conclusion viz. that $r_2 < 1$ is necessary and sufficient for $T_2^m \to 0$.

The significance of Theorem 2.4 is now clear: *either both the Jacobi and Gauss–Seidel methods converge, in which case, since $r_2 < r_1$, the latter converges (asymptotically) more quickly;* or neither method converges, in the sense that $T_i^m \nrightarrow 0$, $i = 1, 2$.

In a more general framework, where T_1 is not necessarily assumed irreducible, this statement, together with the analogue of Theorem 2.4, has come to be known as the Stein–Rosenberg Theorem.

Bibliography and discussion to §2.2

This section barely touches on the discussion of iterative methods of solution of linear equation systems; an exhaustive treatment of this topic with extensive

† Which goes back to Oldenburger (1940); see e.g. also Debreu & Herstein (1953); Varga (1962), Chapter 1.

referencing is given by Varga (1962). The two iterative methods discussed and compared here are more accurately known as the *point* Jacobi and *point* Gauss–Seidel iterative methods; both have various other designations. The proof of (our restricted case of) Theorem 2.4 follows the original paper of Stein & Rosenberg (1948); the proof given in Varga (1962) is different. Generalizations of the Stein–Rosenberg Theorem, which enable one to compare the convergence rates of any two iterative methods out of a class of such methods, are available: see e.g. Theorem 3.15 and subsequent discussion in Varga (1962) where appropriate references are given; and Householder (1958).

Exercises on §2.2

2.7. By considering first, for orientation, the example

$$A = \begin{bmatrix} -1 & 2 & 3 \\ 4 & -5 & 6 \\ 7 & 8 & -9 \end{bmatrix},$$

show that (whether T_1 is irreducible or not) T_2 is reducible.

2.8.† Suppose we are concerned with the iterative solution of the equation system

$$(I - T)\,\boldsymbol{x} = \boldsymbol{b}$$

where $T = \{t_{ij}\}$ is irreducible, and its Perron–Frobenius eigenvalue satisfies $r < 1$. but some of the t_{ii}, $i = 1, \ldots, n$ are positive, although all less than unity. One may then be inclined to make use of this 'natural form' of the matrix $A = I - T$ by putting

$$T = \bar{L} + \bar{U}$$

where $\bar{L}$ consists of the corresponding elements of T below the diagonal, and zeroes elsewhere, and $\bar{U}$ consists of zeroes below the diagonal, but has its other entries, *on the diagonal* as well as above, the same at T. Show that the discussion of the Jacobi and Gauss–Seidel methods, including the relevant part of Theorem 2.4, goes through with $\bar{L}$, $\bar{U}$ replacing L, U respectively.

(Stein & Rosenberg, 1948)

2.9. In the framework of Exercise 2.8, if r_1 and r_2 have the same meaning as in the development of §2.2, show that $r_1 < r$ (so that the 'natural' Jacobi method described in Exercise 2.8 is less efficient than the standard method; an analogous result is true for the 'natural' Gauss–Seidel method as compared with the standard one).

† Exercises 2.8 and 2.9 have a common theme.

2.3 Some extensions of the Perron–Frobenius structure

The theory of non-negative matrices may be used to determine analogous results for certain other matrices which have related structure, and occur in contexts as diverse as mathematical economics and number theory.

The foremost example occurs in the nature of square real matrices $B = \{b_{ij}\}$ $i, j = 1, \ldots, n$ where $b_{ij} \geqslant 0, i \neq j$. We shall call such matrices ML-matrices, and examine their properties.

Closely related properties are clearly possed by matrices which have the form $-B$, where B is an ML-matrix.

Matrices of the type B or $-B$ are sometimes associated with the names of Metzler and Leontief in mathematical economics (the matrix $sI - T$ of §2.1 is of the form $-B$); and under an additional condition of the sort

$$\sum_i b_{ij} \leqslant 0, \text{ all } j; \quad \text{or} \quad \sum_j b_{ij} \leqslant 0, \text{ all } i,$$

with the names of Minkowski and Tambs-Lyche; and with the notion of a transition intensity matrix in the theory of finite Markov processes.

Irreducible ML-matrices

An ML-matrix B may always be related to a non-negative matrix $T \equiv T(\mu)$ through the relation

$$T = \mu I + B$$

where $\mu \geqslant 0$, and is sufficiently large to make T non-negative.

DEFINITION 2.1. An ML-matrix B is said to be irreducible† if T is irreducible.

(This definition is merely a convenience; irreducibility can clearly be defined – equivalently – directly in terms of the non-diagonal elements of B, in terms of 'cycles' – see Definition 1.2.)

By taking μ sufficiently large, the corresponding irreducible T can clearly be made *aperiodic* also and thus primitive; e.g. take $\mu > \max_i |b_{ii}|$.

We confine ourselves to irreducible ML-matrices B in this section, in line with the structure of the whole present chapter. The more important facts concerning such matrices are collected in the following theorem; for the reducible case see the exercises to this section.

THEOREM 2.5. Suppose B is an $(n \times n)$ irreducible ML-matrix. Then there exists an eigenvalue τ such that:

(*a*) τ is real;

(*b*) with τ are associated strictly positive left and right eigenvectors, which are unique to constant multiples;

† In the context of numerical analysis, an irreducible ML-matrix, C, is sometimes called *essentially positive.*

(*c*) $\tau > \text{Re}\,\lambda$ for any eigenvalue λ, $\lambda \neq \tau$, of B (i.e. τ is larger than the real part of any eigenvalue λ of B, $\lambda \neq \tau$);
(*d*) τ is a simple root of the characteristic equation of B;
(*e*) $\tau \leqslant 0$ if and only if there exists $\mathbf{y} \geqslant \mathbf{0}, \neq \mathbf{0}$ such that $B\mathbf{y} \leqslant \mathbf{0}$, in which case $\mathbf{y} > \mathbf{0}$; and $\tau < 0$ if and only if there is inequality in at least one position in $B\mathbf{y} \leqslant \mathbf{0}$;
(*f*) $\tau < 0$ if and only if

$$\Delta_i > 0, i = 1, 2, \ldots, n$$

where Δ_i is the principal minor of $-B$ formed from the first i rows and columns of $-B$;
(*g*) $\tau < 0$ if and only if $-B^{-1} > 0$.

Proof. Writing $B = T - \mu I$ for μ sufficiently large to make T non-negative and irreducible, and noting as a result that if λ_i is an eigenvalue of B, then T has corresponding eigenvalue $\delta_i = \mu + \lambda_i$, and conversely, (*a*), (*b*), (*c*) and (*d*) follow from the Perron–Frobenius theory, τ being identified with $r - \mu$, where r is the Perron–Frobenius eigenvalue of T.

To see the validity of (*e*), let $\lambda_j = x_j + iy_j$ be an eigenvalue of B, $\lambda_j \neq \tau$, and suppose $x_j \geqslant \tau$. If

$$x_j > \tau,\ \delta_j = \mu + \lambda_j = \mu + x_j + iy_j,$$

where
$$\mu + x_j > \mu + \tau > 0,$$
so that
$$|\delta_j| > r = \mu + \tau$$

which is impossible. On the other hand, if $x_j = \tau$, but $y_j \neq 0$, again

$$|\delta_j| > \mu + \tau;$$

so the only possibility is $x_j = \tau$, $y_j = 0$ i.e. $\lambda_j = \tau$, which is again a contradiction.

The condition $B\mathbf{y} \leqslant \mathbf{0}$ may be written as

$$T\mathbf{y} \leqslant \mu\mathbf{y} \qquad (\mu > 0)$$

which is tantamount to $\mu \geqslant r$ i.e.

$$\tau = r - \mu \leqslant 0,$$

by the Subinvariance Theorem (Theorem 1.6). The discussion of strict inequality follows similarly (also in Exercise 2.1).

The validity of (*f*) follows from Theorem 2.2; and of (*g*) from Corollary 1 of Theorem 2.1. #

An ML-matrix, B, occurs frequently in connection with the matrix exp (Bt), $t > 0$, in applications where the matrix exp (Bt), is defined (in analogy to the scalar case) as the pointwise limit of the infinite series† (which converges absolutely pointwise for each $t > 0$):

$$\sum_{k=0}^{\infty} (Bt)^k/k!$$

THEOREM 2.6. An ML-matrix B is irreducible if and only if $\exp(Bt) > 0$ for all $t > 0$. In this case

$$\exp(Bt) = \exp(\tau t)\, wv' + 0(e^{\tau' t})$$

elementwise as $t \to \infty$, where w, v' are the positive right and left eigenvectors of B corresponding to the 'dominant' eigenvalue τ of B, normed so that $v'w = 1$; and $\tau' < \tau$.

Proof. Write $B = T - \mu I$ for sufficiently large $\mu > 0$, so that T is non-negative. Then

$$\begin{aligned}\exp(Bt) &= \exp(-\mu t I)\exp(Tt)\\ &= \exp(-\mu t)\exp(Tt)\end{aligned}$$

and since
$$\exp(Tt) = \sum_{k=0}^{\infty} (tT)^k/k\,!$$

it follows that $\exp(Bt) > 0$ for any $t > 0$ if and only if T is irreducible, which is tantamount to the required.

Suppose now B is irreducible. Then with judicious choice of μ, $T = \mu I + B$ is primitive, with its Perron–Frobenius eigenvalue $r = \mu + \tau$. Invoking Theorem 1.2, we can write

$$T^k = r^k\, wv' + 0\,(k^s\,|\lambda_2|^k),\ |\lambda_2| < r$$

as $k \to \infty$, where w, v' have the properties specified in the statement of the present theorem, since the Perron–Frobenius eigenvector of T correspond to those of B associated with τ.

We may therefore write, for some δ, $0 < \delta < r$ that

$$T^k - r^k\, wv' = Y(k)\,\delta^k$$

where (the elements of) $Y(k) \to 0$ as $k \to \infty$.

Hence
$$\sum_{k=0}^{\infty} \frac{t^k T^k}{k!} - \sum_{k=0}^{\infty} \frac{t^k r^k}{k!} wv' = \sum_{k=0}^{\infty} Y(k)\frac{(t\delta)^k}{k!}$$

i.e. elementwise
$$|\exp(Tt) - \exp(rt)\, wv'| \leqslant \exp(t\delta)\, Y$$

where Y is a matrix of positive elements which bound the corresponding elements of $Y(k)$ in modulus, uniformly for all k. Thus cross multiplying by $\exp(-\mu t)$,

$$|\exp(Bt) - \exp(\tau t)\, w\, v'\,| \leqslant \exp\,[t(\delta - \mu)]\ Y$$

† See Lemma B.2 of Appendix B.

where $\delta - \mu < r - \mu = \tau$. Hence choosing τ' to be any fixed number in the interval $\delta - \mu < \tau' < \tau$, the assertion follows. #

Perron matrices

Irreducible non-negative matrices T, and matrices B and $-B$ where B is an irreducible ML-matrix are special classes of the set of Perron matrices.

DEFINITION 2.2. An $(n \times n)$ matrix $A = \{a_{ij}\}$ is said to be a *Perron matrix* if $f(A) > 0$ for some polynomial f with real coefficients.

Thus if T is non-negative irreducible, an appropriate polynomial, f, is given by

$$f(x) = \sum_{i=1}^{n} x^i,$$

since the matrix

$$f(T) = \sum_{i=1}^{n} T^i$$

will have a strictly positive contribution for any one of its elements from at least one of the T^i, $i = 1, \ldots, n$ (see Exercise 1.3).

An irreducible ML-matrix B may be written in the form $T - \mu I$, $\mu > 0$, where T is non-negative and primitive, so that $(B + \mu I)^k > 0$ for some positive integer k, and this is a real polynomial in B.

A further important subclass of Perron matrices, not previously mentioned, are the *power–positive* matrices A, i.e. matrices A such that $A^k > 0$ for some positive integer k. In this case, clearly, $f(x) = x^k$.

Perron matrices retain some of the features of the Perron–Frobenius structure of irreducible non-negative matrices discussed in Chapter 1, and we discuss these briefly.

THEOREM 2.7. Suppose A is an $(n \times n)$ Perron matrix. Then there exists an eigenvalue τ such that

(*a*) τ is real;
(*b*) with τ can be associated strictly positive left and right eigenvectors, which are unique to constant multiples.
(*c*) τ is a simple root of the characteristic equation of A.

Proof. (*a*) and (*b*): For $\boldsymbol{x} > \boldsymbol{0}$, let

$$\tau(\boldsymbol{x}) = \min_i \sum_j \frac{a_{ij}x_j}{x_i}.$$

Then, since for each i

$$x_i\,\tau(\boldsymbol{x}) \leqslant \sum_j a_{ij}\,x_j$$

$$\boldsymbol{x}\,\tau(\boldsymbol{x}) \leqslant A\boldsymbol{x}$$

so that $$\mathbf{1}'x\,\tau(x) \leqslant \mathbf{1}'A\,x$$

i.e. $$\tau(x) \leqslant \mathbf{1}'Ax/\mathbf{1}'x \leqslant K.$$

since each element of $\mathbf{1}'A$ is less than or equal to $\max_i \sum_i a_{ij} \equiv K$, so that $\tau(x)$ is bounded above *uniformly* for all $x > 0$. Now, define

$$\tau = \sup_{x>0} \tau(x).$$

It follows that $\tau \geqslant \tau(1) = \min_i \sum_{j=1}^{n} a_{ij}$.

Now, let τ^* be defined by

$$\tau^* = \sup_{z \epsilon \mathscr{C}} \tau(z)$$

where $\mathscr{C}$ is the set of vectors $z = f(A)y$ where $y \geqslant \mathbf{0}, y'y = 1$. Since this set of ys is compact, and the mapping is continuous on the set, $\mathscr{C}$ is compact, and since $z \,\epsilon \mathscr{C} \Rightarrow z > \mathbf{0}$, clearly $\mathscr{C}$ is a subset of the set $\{x; x > \mathbf{0}\}$. Thus

$$\tau^* \leqslant \tau \tag{2.8}$$

and since $\tau(x)$ is a continuous mapping from $\{x; x > 0\}$ to R_1, it follows that τ^* is attained for some $z^* \,\epsilon \mathscr{C}$. Now, since for any $x > 0$

$$Ax - x\tau(x) \geqslant \mathbf{0}$$

and x may be taken as normed to satisfy $x'x = 1$, (without change in $\tau(x)$) if we multiply this inequality from the left by $f(A)$

$$A(f(A)x) - (f(A)x)\,\tau(x) \geqslant \mathbf{0}$$

since $f(A)A = Af(A)$; and since $w = f(A)x \,\epsilon \mathscr{C}$ it follows that

$$\tau(w) \geqslant \tau(x)$$

so that certainly

$$\tau^* \geqslant \tau$$

whence $$\tau^* = \tau, \text{ from (2.8).}$$

Thus we have that for some $z^* > \mathbf{0}$

$$Az^* - \tau z^* \geqslant \mathbf{0}.$$

Suppose now that the inequality is strict in at least one position. Then, as before

$$A(f(A)z^*) - \tau(f(A)z^*) > \mathbf{0}$$

so that for a $w = f(A)z^* > 0$

$$\tau(w) = \min_i \sum_j \frac{a_{ij}w_j}{w_i} > \tau$$

for each i, and this is a contradiction to the definition of τ. Hence

$$Az^* = \tau z^*,$$

so that, to complete the proof of assertions (*a*) and (*b*) of the theorem, it remains first to prove uniqueness to constant multiples of the eigenvector corresponding to τ; and secondly to prove that there is a similar left eigenvector structure.

Since, for arbitrary positive integer k, and *any eigenvector* (possibly with complex elements), $\boldsymbol{b}$

$$A^k \boldsymbol{b} = \tau^k \boldsymbol{b}$$

it follows that

$$f(A)\boldsymbol{b} = f(\tau)\boldsymbol{b}.$$

Now, $f(A)$ is a positive matrix, and since for $\boldsymbol{b}$ we can put $z^*, > \boldsymbol{0}$, it follows from the Subinvariance Theorem that $f(\tau)$ is the Perron–Frobenius eigenvalue of $f(A)$ and hence $\boldsymbol{b}$ must be a multiple of z^*, by the Perron–Frobenius Theorem.

As regards the left eigenvector structure, it follows that all the above theory will follow through, *mutatis mutandis,* in terms of a real number τ' replacing τ, where

$$\tau' = \sup_{x>0} \left\{ \min_j \frac{\sum_{i=1}^{n} x_i a_{ij}}{x_j} \right\}$$

so that there exists a unique (to constant multiples) positive left eigenvector $\boldsymbol{c}$ satisfying

$$\boldsymbol{c}'A = \tau'\boldsymbol{c}'$$

and also

$$Az^* = \tau z^*.$$

Hence $\qquad \boldsymbol{c}'Az^* = \tau'\boldsymbol{c}'z^* = \tau\boldsymbol{c}'z^*$, and since $\boldsymbol{c}'z^* > 0$, $\tau' = \tau$.

(*c*): The fact that there is (to constant multiples) only one real $\boldsymbol{b} \neq \boldsymbol{0}$ such that

$$(\tau I - A)\boldsymbol{b} = \boldsymbol{0}$$

implies that $(\tau I - A)$ is of rank precisely $n-1$, so that some set of $(n-1)$ of its columns is linearly independent. Considering the $(n-1) \times n$ matrix formed from these $(n-1)$ columns, since it is of full rank $(n-1)$, it follows† that $(n-1)$ of its rows must be linearly independent. Thus by crossing out a certain row and a certain column of $(\tau I - A)$ it is possible to form an $(n-1) \times (n-1)$ non-singular matrix, whose determinant is therefore non-zero. Thus

$$\text{Adj}\,(\tau I - A) \neq 0$$

† By a well-known result of linear algebra.

since at least one of its elements is non-zero. Thus, as in the proof of the Perron–Frobenius Theorem (part (*f*)), Adj $(\tau I - A)$ has all its elements non-zero, and all positive or all negative, since all columns are real non-zero multiples of the positive right eigenvector, and the rows are real non-zero multiples of the positive left eigenvector, the sign of one element of Adj $(\tau I - A)$ determining the sign of the whole.

Hence
$$\phi'(\tau) \neq 0,$$
as in the proof of Theorem 1.1. #

COROLLARY 1.

$$\min_i \sum_j a_{ij} \leqslant \tau \leqslant \max_i \sum_j a_{ij}$$

with a similar result for the columns.

Proof. We proved in the course of the development, that

$$\min_i \sum_j a_{ij} \leqslant \tau \leqslant \max_j \sum_i a_{ij}$$

and implicitly (through considering τ') that

$$\min_j \sum_i a_{ij} \leqslant \tau \leqslant \max_i \sum_j a_{ij}.$$

COROLLARY 2. Either Adj $(\tau I - A) > 0$ or $-$Adj $(\tau I - A) > 0$. (Proved in the course of (*c*)).

COROLLARY 3. (Subinvariance). For some real s,

$$\boldsymbol{Ax} \leqslant s\boldsymbol{x}$$

for some $\boldsymbol{x} \geqslant \boldsymbol{0}, \neq \boldsymbol{0}$ implies $s \geqslant \tau$; $s = \tau$ if and only if $\boldsymbol{Ax} = s\boldsymbol{x}$.

Proof. Let $s\boldsymbol{x} - \boldsymbol{Ax} = \boldsymbol{f}; \boldsymbol{f} \geqslant \boldsymbol{0}$.

Then
$$s\,\boldsymbol{c}'\boldsymbol{x} - \boldsymbol{c}'\boldsymbol{Ax} = \boldsymbol{c}'\,\boldsymbol{f},$$
i.e.
$$s\,\boldsymbol{c}'\boldsymbol{x} - \tau\boldsymbol{c}'\boldsymbol{x} = \boldsymbol{c}'\boldsymbol{f}.$$

Hence, since $s - \tau = \boldsymbol{c}'\boldsymbol{f}/\boldsymbol{c}'\boldsymbol{x}$, the result follows. #

Since the class of Perron matrices includes (irreducible) periodic non-negative matrices, it follows that there may be other eigenvalues of A which have modulus equal to that of τ. This cannot occur for *power-positive* matrices A, which are clearly a generalization of primitive, rather than irreducible, non-negative matrices.

For suppose that λ is any eigenvalue of the power-positive matrix A; thus for some possibly complex valued $\boldsymbol{b} \neq \boldsymbol{0}$,

$$\boldsymbol{Ab} = \lambda\boldsymbol{b}$$

and so
$$\boldsymbol{A}^k\boldsymbol{b} = \lambda^k\boldsymbol{b}$$

for each positive integer k, and hence for that k for which $A^k > 0$. It follows as before that for this k, τ^k is the Perron–Frobenius eigenvalue of the primitive matrix A^k, and hence is uniquely dominant in modulus over all other eigenvalues.

Hence, in particular, for this k

$$|\tau^k| > |\lambda^k| \qquad \text{i.e. } |\tau| > \lambda$$

if $\lambda^k \neq \tau^k$.

Now suppose $\lambda^k = \tau^k$. Then it follows that any eigenvector $\boldsymbol{b}$ corresponding to λ must be a multiple of a single *positive* vector, since A^k is primitive, corresponding to the Perron–Frobenius eigenvalue τ^k of A^k.

On the other hand this positive vector itself must correspond to the eigenvalue τ of A *itself*; call it z^*.

Thus $$Az^* = \tau z^*, Az^* = \lambda z^*$$

Hence $$c'Az^* = \tau c'z^* = \lambda c'z^*$$

Therefore $$\tau = \lambda.$$

In this situation also, τ is clearly a *uniquely dominant eigenvalue in modulus* as with primitive matrices. It may, however, be *negative since for even k,* $A^k > 0$ where $A = -T$, $T > 0$.

Bibliography and discussion to §2.3

The theory of ML-matrices follows readily from the theory developed earlier for non-negative matrices T and the matrices $sI - T$ derived from them, and no separate discussion is really necessary. Theorem 2.6 is useful, apart from other contexts, in the theory of Markov processes on finite state space, in that it describes the asympototic behaviour of a probability transition (sub–) matrix $\exp(Bt)$ at time t, as t becomes large (B in this context being irreducible), and satisfying, for all i, $\sum_j b_{ij} \leqslant 0$ (see for example Mandl (1960); Darroch & Seneta (1967)).

A question closely related to one discussed for ML-matrices is that of matrices whose inverses exist and have non-negative real entries. Ostrowski (1937; 1956) calls a real ($n \times n$) matrix $A = \{a_{ij}\}$ an M-matrix if: $a_{ij} \leqslant 0$, $i \neq j$, and A^{-1} exists, with $A^{-1} \geqslant 0$, in effect, and deduces certain results for such matrices. We shall not pursue these further in this book, mentioning only the book of Varga (1962) and the papers of Fan (1958) and Fiedler & Pták (1962), for further discussion and references.

The notion of a Perron matrix seems to be due to Dionísio (1963/4, Fasc. 1) who gives the results of Theorem 2.7, his method of proof (as also ours) imitating that of Wielandt (1950) as modified in Gantmacher (1959). It will be noted that this approach differs somewhat from that used in the proof of the Perron–Frobenius Theorem itself in our Chapter 1, that proof following Wielandt almost exclusively. Both proofs have been presented in the text for interest, although the initial stages of Wielandt's proof are not quite appropriate in the general context of a Perron matrix.

Power-positive matrices appear to have been introduced by Brauer (1961), whose paper the interested reader should consult for further details.

There are various other (finite–dimensional) extensions of the Perron–Frobenius structure of non-negative matrices. One such which has received substantial attention is that of an operator H, not necessarily linear, mapping $\{x; x \geqslant 0\}$ into itself, which is *monotone* ($0 \leqslant x_1 \leqslant x_2 \Rightarrow Hx_1 \leqslant Hx_2$), *homogeneous* ($H(\alpha x) = \alpha H(x)$ for $0 \leqslant \alpha < \infty$), and *continuous*, these properties being possessed by any non-negative matrix T. A weak analogue of the irreducibility is

$$\{0 \leqslant x_1 \leqslant x_2, x_1 \neq x_2\} \Rightarrow \{Hx_1 \leqslant Hx_2, Hx_1 \neq Hx_2\};$$

and for primitivity the right hand side of the implication must be supplemented by the existence of a positive integer in such that $H^m x_1 < H^m x_2$. A theorem much like the Perron–Frobenius Theorem in several respects can then be developed using a method of proof rather similar to that in Chapter 1. The interested reader should consult Brualdi, Parter & Schneider (1966), Morishima (1961, 1964) and Solow & Samuelson (1953).

It is also relevant to mention the paper of Mangasarian (1971) who considers the generalized eigenvalue problem $(A - \lambda B)x$ under variants of the assumption that $y'B \geqslant 0', \neq 0' \Rightarrow y'A \geqslant 0', \neq 0'$ to derive a Perron–Frobenius structure; the nature of this generalization becomes clear by putting $B = I$.

Finally we note certain generalizations of Perron–Frobenius structure obtainable by considering matrices which are not necessarily non-negative, but specified *cycles* of whose elements are (the definition of *cycle* is as in §1.2). This kind of theory has been developed in e.g. Maybee (1967); Bassett, Maybee & Quirk (1968); and Maybee & Quirk (1969) – from which further references are obtainable. A central theme of this work is the notion of a *Morishima matrix* (Morishima, 1952), which has properties closely allied to the Perron–Frobenius structure.

An $(n \times n)$ non-negative matrix T, with a view to further generalizations, may itself be regarded as a matrix representation of a linear operator which maps a convex cone of a partially ordered vector space (here the positive orthant of R_n) into itself, in the case when the cone is determined by its set of extreme generators.

Exercises on §2.3

2.10. Suppose B is an arbitrary $(n \times n)$ ML-matrix. Show that there exists a real eigenvalue ρ^* such that $\rho^* \geqslant \text{Re}\,\lambda$ for any eigenvalue λ of B, and that $\rho^* < 0$ if and only if $\Delta_i > 0$, $i = 1, 2, \ldots, n$. (See Exercises to §2.1 of this chapter.)

2.11. In the situation of Exercise 2.10, show that if $\rho^* \leqslant 0$ then $\Delta_i \geqslant 0$, $i = 1, 2, \ldots, n$.

2.12. Let $B = \{b_{ij}\}$ be an $(n \times n)$ ML-matrix with $b_{ij} > 0, i \neq j$, whose 'dominant' eigenvalue is denoted by ρ^*. Show that (for $n \geqslant 2$) the condition: $\Delta_n \geqslant 0$, $\Delta_i > 0$, $i = 1, 2, \ldots, n-1$, ensures that $\rho^* \leqslant 0$.

Hint: Make use of the identity $(-B)\operatorname{Adj}(-B) = \Delta_n I$, and follow the proof pattern of the latter part of Theorem 2.2.

[Kotelyanskii (1952) shows that for such B the apparently weaker condition $\Delta_i \geqslant 0$, $i = 1, \ldots, n$ *implies* $\Delta_i > 0$, $i = 1, \ldots, n-1$. Thus, taking into account also Exercise 2.11 above, $\Delta_i \geqslant 0$, $i, \ldots, n$ is necessary and sufficient for $\rho^* \leqslant 0$ for such B.]

2.13. Let $B = \{b_{ij}\}$ be an irreducible $(n \times n)$ ML-matrix with 'dominant' eigenvalue ρ^*. Show that

$$\min_i \sum_{j=1}^n b_{ij} \leqslant \rho^* \leqslant \max_i \sum_{j=1}^n b_{ij}$$

with either equality holding if and only if both hold, making use of the analogous result for a non-negative irreducible matrix T.

If it is assumed, further, that

$$\sum_{j=1}^n b_{ij} \leqslant 0$$

for every i, show that: $\Delta_i \geqslant 0$, $i = 1, \ldots, n$; and that $\Delta_i > 0$, $i = 1, \ldots, n$ if and only if some row sum of B is *strictly* negative. Hence deduce that $\Delta_n \neq 0$ if and only if some row sum of B is strictly negative. [The variants of this last statement, have a long history; see Taussky (1949).]

2.14. If B is an $(n \times n)$ ML-matrix, not necessarily irreducible, but satisfying

$$\sum_{j=1}^n b_{ij} \leqslant 0,$$

show that $\Delta_i \geqslant 0$, $i = 1, \ldots, n$. (Ledermann, 1950*a*)

2.15. Show that a non-negative reducible matrix T cannot be a Perron matrix.

2.4 Combinatorial properties

It has been remarked several times already in Chapter 1 that many properties of a non-negative matrix T depend only on the *positions* of the positive and zero elements within the matrix, and *not on the actual size* of the positive elements. Thus the classification of indices into essential and inessential, values of periods of indices which communicate with themselves, and hence investigation of the properties of irreducibility and primitivity in relation to a given non-negative T, all depend only on the location of the positive entries. This is a consequence of the more general fact that the positions of the positive and zero elements in all powers, T^k, k a positive integer, depend only on the positions in T.

It thus follows that to make a general study of the sequence of the powers T^k, $k = 1, 2, \ldots$ it suffices for example to replace, for each $k \geq 1$, in the matrix T^k the element $t_{ij}^{(k)}$ by unity, whenever $t_{ij}^{(k)} > 0$. The matrices so obtained may still be obtained as appropriate powers of the matrix T so modified (denoted say by $\tilde{T}$), if we accept that the rules of addition and multiplication involving elements 0, 1 pairwise are those of a simple Boolean algebra, i.e. the rules are the usual ones, with the single modification that $1 + 1 = 1$. The basic matrix $\tilde{T}$ has been called the incidence matrix of the matrix T in Chapter 1; such matrices with the Boolean algebra rules for composition as regards elements of their products, are also known in certain contexts as *Boolean relation matrices.*

An alternative formulation for the study of the power structure of a non-negative matrix is in graph theoretic terms; while a third is in terms of a mapping, F, induced by a matrix T, of its index set $\{1, 2, \ldots, n\}$ into itself. We shall pursue this last approach since it would seem to have the advantage of conceptual simplicity (as compared to the graph-theoretic approach) within the limited framework with which we are concerned in the present chapter. Nevertheless, it should be stressed that all three approaches are quite equivalent, merely being different frameworks for the same topic, viz. the *combinatorial* properties of non-negative matrices.

Denote now by S the set of indices $\{1, 2, \ldots, n\}$ of an irreducible matrix T, and let $L \subset S$. Further, for integer $h \geq 0$, let $F^h(L)$ be the set of indices $j \in S$ such that

$$t_{ij}^{(h)} > 0 \text{ for some } i \in L.$$

(If $L = \phi$, the empty set, put $F^h(\phi) = \phi$.)

Thus $F^h(i)$ is the set of $j \in S$ such that $t_{ij}^{(h)} > 0$. Also $F^0(L) = L$ by convention. Further, we notice the following easy consequences of these definitions:

(i) $A \subset B \subset S$, then $F(A) \subset F(B)$.
(ii) If $A \subset S$, $B \subset S$, then $F(A \cup B) = F(A) \cup F(B)$.
(iii) For integer $h \geq 0$, and $L \subset S$,

$$F^{h+1}(L) = F(F^h(L)) = F^h(F(L)).$$

(iv) The mapping F^h may be interpreted as the F-mapping associated with the non-negative matrix T^h, or as the hth iterate (in view of (iii)) of the mapping F associated with the matrix T.

We shall henceforth restrict outselves to irreducible matrices T; and eventually to primitive matrices T, in connection with a study of the *index of primitivity* of such T.

DEFINITION 2.3. The minimum positive integer $\gamma \equiv \gamma(T)$ for which a primitive matrix T satisfies $T^\gamma > 0$ is called the index of primitivity (or exponent) of T.

Irreducible matrices

We shall at the outset assume that $n \geqslant 2$, to avoid trivialities, and note that for an irreducible T, $F(i) \neq \phi$, for each $i \in S$.

LEMMA 2.1. $F(S) = S$. If L is a proper subset of S *and* $L \neq \phi$ then $F(L)$ contains some index not in L.

Proof. If $F(S)$ were a proper subset of S, then T would contain a zero column; this is not possible for irreducible T. If for a non-empty proper subset L of S, $F(L) \subset L$, then this also contradicts irreducibility, since then $F^h(L) \subset L$ all positive integer h, and hence for $i \in L$, $j \notin L$, $i \not\rightarrow j$. #

LEMMA 2.2. For $0 \leqslant h \leqslant n-1$, $\{i\} \cup F(i) \cup \ldots \cup F^h(i)$ contains at least $h+1$ indices.

Proof. The proposition is evidently true for $h = 0$. Assume it is true for some h, $0 \leqslant h < n-1$; then

$$L = \{i\} \cup F(i) \cup \ldots \cup F^h(i)$$

contains at least $h+1$ indices, and one of two situations occurs:

(a) $L = S$, in which case

$$\{i\} \cup F(i) \cup \ldots \cup F^{h+1}(i) = S$$

also, containing $n > h+1$ elements, so that $n \geqslant h+2$, and the hypothesis is verified; or

(b) L is a proper non-empty subset of S in which case

$$F(L) = F(i) \cup \ldots \cup F^{h+1}(i)$$

contains at least one index not in L (by Lemma 2.1), and since $i \in L$,

$$\{i\} \cup F(L) = \{i\} \cup F(i) \cup \ldots \cup F^{h+1}(i)$$

contains all the indices of L and at least one not in L, thus containing at least $h+2$ elements. #

COROLLARY 1. If $\{i\} \cup F(i) \cup \ldots \cup F^{h-1}(i)$ is a proper subset of S, then, with the union of $F^h(i)$, at least one new element is added. Thus, *if*

$$\{i\} \cup F(i) \cup \ldots \cup F^h(i), \qquad h \leqslant n-2,$$

contains precisely $h +$ [illegible]s, then union with each successive $F^r(i)$, $r = 1, 2, \ldots, n-2$ adds [illegible]*sely one new element.*

COROLLARY 2. For any $i \in S$, $i \cup F(i) \cup \ldots \cup F^{n-1}(i) = S$. (This was proved directly in Exercise 1.3.)

For purposes of the sequel we introduce a new term.

DEFINITION 2.4. An irreducible $(n \times n)$ matrix T is said to be *deterministic* if it is periodic with period $d = n$ (i.e. each cyclic subset of the matrix contains only one index). Equivalently, for such an irreducible matrix, for each $i \in S$ there is only one $j \in S$ such that $t_{ij} > 0$.)

LEMMA 2.3. For $i \in S$, $F^h(i)$ contains at least two indices for some h, $1 \leqslant h \leqslant n$ unless the deterministic case obtains.

Proof. Since $\{i\} \cup F(i) \cup \ldots \cup F^{n-1}(i) = S$, two cases are possible:

(*a*) each of the $F^h(i)$, $h = 0, \ldots, n-1$, contains precisely one index, and all differ from each other. Now if $F^n(i)$ contains only one index, either $F^n(i) = \{i\}$ and we are in the deterministic case; or $F^n(i) = F^h(i)$ for some h, $1 \leqslant h \leqslant n-1$, which is impossible as irreducibility of T is contradicted. Otherwise $F^n(i)$ contains at least two indices.

(*b*) some $F^h(i)$, $1 \leqslant h \leqslant n-1$ contains at least two indices. #

We now pass on to a study of a general upper bound for $\gamma(T)$ *depending only on the dimension n* ($\geqslant 2$ by earlier assumption) for primitive T. For subclasses of $(n \times n)$ primitive matrices T satisfying additional structural conditions, stronger results are possible.†

THEOREM 2.8. For a primitive $(n \times n)$ matrix T, $\gamma \leqslant n^2 - 2n + 2$.

Proof. According to Lemma 2.3, for arbitrary fixed $i \in S$,

$$\{i\} \cup F(i) \cup \ldots \cup F^{n-1}(i) = S,$$

and either (*a*) $F^h(i)$, $h = 0, \ldots, n-1$ all contain precisely one index, in which case $F^n(i)$ contains at least two‡ indices; or (*b*) one of the $F^h(i)$, $1 \leqslant h \leqslant n-1$ contains at least two.

(*a*) Since $\{i\} \cup \ldots \cup F^{n-1}(i) = S$, it follows $F(i) \cup \ldots \cup F^n(i) = F(S) = S$, in which case $F^n(i)$ must contain i, and at least one index not i, i.e. one of $F^h(i)$, $h = 1, \ldots, n-1$. Hence for some integer $m \equiv m(i)$, $1 \leqslant m < n$,

$$F^m(i) \subseteq F^n(i) = F^{m+(n-m)}(i).$$

so that operating repeatedly with F^{n-m}:

$$F^m(i) \subseteq F^{m+(n-m)}(i) \subseteq F^{m+2(n-m)}(i) \subseteq \ldots \subseteq F^{m+(n-1)(n-m)}(i)$$

and by Corollary 2 of Lemma 2.2,

$$F^{m+(n-1)(n-m)}(i) = S.$$

Now $m + (n-1)(n-m) = n + (n-2)(n-m) \leqslant n + (n-2)(n-1)$
$= n^2 - 2n + 2.$

† See Exercise 1.3, and Exercises 2.16 and 2.17 of the sequel and Bibliography and discussion.

‡ The deterministic case is excluded by assumption of primitivity of T.

(*b*) If one of the $F^h(i)$, $h = 1, 2, \ldots, n-1$ contains at least two indices, we further differentiate between two cases:

(*b*.1) $\{i\} \cup F(i) \cup \ldots \cup F^{n-2}(i) \neq S$. Then by Corollary 1 of Lemma 2.2, each of $F^h(i)$, $h = 0, \ldots, n-2$ adds precisely one new element, and by Corollary, 2 $F^{n-1}(i)$ contributes the last element required to make up S. Let $p \equiv p(i)$, $1 \leqslant p \leqslant n-1$, be the smallest positive integer such that $F^p(i)$ contains at least two elements. Then there exists an integer m, $0 \leqslant m < p$ such that $F^m(i) \subseteq F^p(i)$. Proceeding as in (*a*),

$$F^{m+(n-1)(p-m)}(i) = S$$

and

$$m + (n-1)(p-m) = p + (n-2)(p-m) \leqslant (n-1) + (n-2)(n-1)$$
$$= (n-1)^2 < n^2 - 2n + 2.$$

(*b*.2) $\{i\} \cup F(i) \cup \ldots \cup F^{n-2}(i) = S$. Then

$$S = F(S) = F(i) \cup \ldots \cup F^{n-1}(i),$$

as before, so that for some p, $1 \leqslant p \leqslant n-1$, $F^p(i) \supseteq F^0(i) = \{i\}$. Proceeding as before,

$$F^{0+(n-1)p}(i) = S,$$

with

$$(n-1)p \leqslant (n-1)^2 < n^2 - 2n + 2.$$

Thus combining (*a*) and (*b*), we have that for *each* $i \in S$,

$$F^{n^2-2n+2}(i) = S$$

which proves the theorem. #

COROLLARY. $\gamma(T) = n^2 - 2n + 2$ if and only if a simultaneous permutation of rows and columns of T reduces it to an (almost deterministic) form having incidence matrix

$$\tilde{T} = \begin{bmatrix} 0 & 1 & 0 & 0 & \ldots & 0 \\ 0 & 0 & 1 & 0 & \ldots & 0 \\ & & & & \vdots & \\ 0 & 0 & 0 & & \ldots & 1 \\ 1 & 1 & 0 & 0 & \ldots & 0 \end{bmatrix}$$

Proof. From the proof of Theorem 2.8, it may be seen that only in case (*a*) is it possible that there exists a row, say the ith, which becomes entirely positive for the first time only with a power as high as $n^2 - 2n + 2$, and then only with the conditions that $m = 1$ is the unique choice *and*

$$F(i) \subseteq F^n(i) \subseteq \ldots \subseteq F^{1+(n-2)(n-1)}(i) \neq S. \tag{2.9}$$

Ignoring (2.9) for the moment, the other conditions indicate that, taking i as the first index, $F(i)$ as the second, . . ., $F^{n-1}(i)$ as the nth, a form with the incidence matrix displayed is the only one possible. It is then easily checked for this form that (2.9) holds, as required, for *each* successive operation on $F(i)$ by F^{n-1} adds *precisely one new* element, up to the $(n-1)$th operation. #

Bibliography and discussion to §2.4

The approach of the present section is a combination of those of Holladay & Varga (1958) and Pták (1958). Wielandt (1950) stated the result of Theorem 2.8 (without proof) and gave the case discussed in its Corollary as an example that the upper bound is attainable. Theorem 2.8 was subsequently proved by Rosenblatt (1957) using a graph-theoretic approach, and by Holladay & Varga and Pták, in the papers cited. The full force of the Corollary to Theorem 2.8 would seem to be due Dionísio (1963/4, Fasc. 2), using a graph-theoretic approach.

The study of the combinatorial properties of non-negative matrices has expanded rapidly since the early contributions of Rosenblatt, Holladay, Varga, and Pták, usually with the aid of methods which are explicitly graph-theoretic. We shall mention only a limited number of these contributions, from which (or from whose authors) the interested reader should be able to trace others. Perkins (1961), Dulmage & Mendelsohn (1962, 1964) and Heap & Lynn (1964) have further pursued the structure of the powers of primitive T, obtaining sharper bounds than Wielandt's under certain structural conditions. Subsequently, Heap & Lynn (1966, I), have considered the 'oscillatory' structure of the sequence of powers of irreducible and reducible T; and (1966, II) the maximum number of positive elements, ρ, which may be contained by a matrix of the sequence T^r, $r = 1, 2, \ldots$ and the positive integer ν for which T^r first contains ρ positive elements. Schwartz (1967) has made a study similar to that of Heap & Lynn (1966, I) for irreducible matrices, using a combinatorial approach (similar to our discussion).

Exercises on §2.4

All exercises refer to an $(n \times n)$ *primitive* matrix T, unless otherwise stated.

2.16. In Exercise 1.3, of Chapter 1, it was shown that if $t_{ii} > 0$, $i = 1, \ldots, n$, then $\gamma(T) \leqslant n-1$. Adapting the reasoning of this exercise show that

(i) If T has exactly one positive diagonal entry, then $\gamma(T) \leqslant 2(n-1)$.
(Rosenblatt; 1957)

(ii) If T has exactly $r \geqslant 1$ diagonal entries then $\gamma(T) \leqslant 2n - r - 1$.
(Holladay & Varga, 1958)

2.17. If T is an irreducible matrix which has $t_{ij} > 0$ if and only if $t_{ji} > 0$, show that T^2 has *all* its diagonal elements positive. If, an addition, T is primitive, show that $\gamma(T) \leqslant 2(n-1)$.
(Holladay & Varga, 1958)

2.18. Use the fact that $t_{11}^{(r)} > 0$ for some r, $1 \leqslant r \leqslant n$, and the essence of part (i) of Exercise 2.16 above to obtain the bound $\gamma(T) \leqslant 2n^2 - 2n$ for an *arbitrary* (primitive) T. (This bound, substantially weaker than that of Wielandt in Theorem 2.8, was already known to Frobenius (1912)).

2.19. If $k(T)$ is the *total* number of positive entries in T, show that $k(T^r)$, $r = 1, 2, \ldots$ may not (initially) be non-decreasing, by investigating the (9×9) matrix whose positive entries are $t_{11}, t_{12}, t_{23}, t_{24}, t_{25}, t_{36}, t_{37}, t_{38}, t_{46}, t_{47}, t_{48}, t_{56}, t_{57}, t_{58}, t_{69}, t_{79}, t_{89}, t_{91}$. (Šidák, 1964b)

2.20. In the framework of Exercise 2.19, show that $k(T^r)$ is non-decreasing with r if at least $(n - 1)$ diagonal entries of T are positive. (Use Lemma 2.1.) (Šidák, 1964*b*)

2.5 A glance at other topics

In this section we merely mention certain topics which have received a substantial amount of attention in the literature, together with guides to where they may be pursued further.

1. *Inclusion theorems for the Perron–Frobenius eigenvalue*

If T is an irreducible non-negative matrix with Perron–Frobenius eigenvalue r, then in Corollary 1 to Theorem 1.1 of Chapter 1 we obtained the Frobenius-type inclusion

$$s \leqslant r \leqslant S$$

where

$$s = \min_i \sum_{j=1}^{n} t_{ij}, \quad S = \max_i \sum_{j=1}^{n} t_{ij}$$

with either equality holdings if and only if all row sums are equal. Related inequalities were discussed in Exercises 1.6 to 1.8 of Chapter 1.

This leads to the problem, in the case that not all row sums are the same, of determining positive numbers p_1 and p_2 such that

$$s + p_2 \leqslant r \leqslant S - p_1$$

suggested by Ledermann in 1950. For positive T, such numbers p_1 and p_2, of successively 'sharper' nature, were obtained by Ledermann (1950*b*), Ostrowski (1952), and Brauer (1957*a*). The case of irreducible T was similarly considered by Ostrowski & Schneider (1960). An account of these and other contributions is given by Marcus & Minc (1964).

2. *Localization theorems for the spectrum*

The first problem which is of interest here, posed by Kolmogorov in 1938, is that of characterizing the region M_n of the complex plane which consists of all points which can be eigenvalues of $(n \times n)$ non-negative matrices T; clearly for the problem to be meaningful it is necessary to impose a condition restricting the size of the dominant non-negative eigenvalue, ρ. In fact it is, as can be shown by a simple similarity transformation argument of Kolmogorov (see e.g. Gantmacher (1959, p. 105)), adequate to discuss the problem for $(n \times n)$ matrices T which are *stochastic,* if not only n, but also ρ, is fixed.

DEFINITION 2.5. A non-negative matrix $T = \{t_{ij}\}$, $i, j = 1, 2, \ldots, n$ is called stochastic† (or, more precisely row stochastic) if

$$\sum_{j=1}^{n} t_{ij} = 1, \text{ each } i.$$

It is called doubly stochastic if also

$$\sum_{i=1}^{n} t_{ij} = 1, \text{ each } j.$$

It is clear that since all row sums of a stochastic matrix have equal value, viz. unity, that $\rho = 1$, and so, for any eigenvalue λ of a stochastic matrix

$$|\lambda| \leqslant 1.$$

Thus clearly for any fixed n, and stochastic T, M_n is contained in the unit circle.

This problem was partially solved by Dmitriev & Dynkin (1945; Dmitriev, (1946), and the solution completed by Karpelevich (1951). It turns out that M_n consists of the interior and boundary of a simple curvilinear polygon with vertices on the unit circle, and is symmetric about the real axis. A description of this and related theory is available in the book of de Oliveira (1968, Chapter 2).

The second problem is that of determining, for a given non-negative T, a region which contains all the eigenvalues λ of T. Theorems which determine regions containing all eigenvalues for general (possibly complex valued) matrices in terms of their elements may be used for this purpose. The field here is broad; an account is available in the books of Varga (1962, § 1.4 and its discussion) and Marcus & Minc (1964); and the recent paper of Timan (1972).

Further, we mention the difficult problem suggested by Suleĭmanova (1949, 1953) of investigating the n-dimensional region $\mathfrak{M}_n$ consisting of n-tuples of (real or complex) numbers, these n-tuples being the set of characteristic roots of some stochastic non-negative matrix T, this being a generalization of Kolmogorov's problem. Associated with this is the problem of conditions on a set of n numbers in order that these form a set of eigenvalues of an $(n \times n)$ stochastic matrix. After Suieĭmanova (1949), this work was carried forward by Perfect (1952, 1953, 1955) for stochastic matrices, and by Perfect &

† For a useful generalization of the notion of stochasticity and some results relevant to this section, see Haynsworth (1955).

Mirsky (1965, and its references) for doubly stochastic matrices. Further work in this direction has been carried out by de Oliveira (1968), whose book contains an extensive description and references.

Vere-Jones (1971) has recently considered another version of this problem, viz. the investigation of the n-dimensional region consisting of points which are eigenvalue sets of those non-negative T which can be diagonalized by a similarity transformation $Q^{-1}TQ$ by a fixed matrix Q. (See also Buharaev (1968).)

We finally mention the articles of Mirsky (1963, 1964) as sources of further information.

3. *Doubly stochastic matrices*

Various properties of doubly stochastic matrices, apart from spectral ones discussed above, have received a great deal of attention in recent years.

One of the founding papers in this direction was that of Birkhoff (1946) who showed that every doubly stochastic matrix can be expressed as a convex combination of permutation matrices.

Another was that of van der Waerden (1926), who conjectured that: if $A = \{a_{ij}\}$ is an $(n \times n)$ doubly stochastic matrix and per(A) denotes the *permanent* of A, i.e.

$$\text{per}\,(A) = \sum_{\sigma} \prod_{i=1}^{n} a_{i,\sigma(i)}$$

where the sum is taken over all permutations σ of $1, \ldots, n$, then

$$\text{per}\,(A) \geqslant n!/n^n$$

with equality holding if and only if $A = J_n$, the matrix all of whose entries are $1/n$. This conjecture and general properties of permanents has been investigated extensively especially by Marcus, Minc and co-workers. References may be found in Marcus & Minc (1964, 1967).

Finally we mention a line of investigation initiated by Sinkhorn (1964), who proved the following theorem:

Let T be a positive square matrix. Then there exist two diagonal matrices D_1, D_2 whose diagonal elements are positive such that $D_1\,T\,D_2$ is doubly stochastic. Moreover these two matrices are uniquely determined up to scalar factors.

The complete extension of this theorem to a T which may contain some zero elements is non-trivial. A sufficient condition on such T for the theorem to hold is that T be irreducible and have positive main diagonal (Brualdi, Parter & Schneider (1966); Djoković (1970)). An analysis of necessary and sufficient conditions may be found in the first of these papers, but particularly in Sinkhorn & Knopp (1967); these two papers should also be consulted for further references. A key idea in this work is that of *full indecomposability* of a non-negative square matrix T:

DEFINITION 2.6. T is fully indecomposable if it is not possible to find permutation matrices P and Q such that

$$PTQ = \begin{bmatrix} B & 0 \\ C & D \end{bmatrix}$$

where B and D are square submatrices.

Clearly, this definition is more restrictive than that of irreducibility (= indecomposability), since irreducibility may be defined in the same way, but with P^T, the transpose of P, replacing Q. Thus a fully indecomposable matrix is irreducible.

3 Some theory and applications of sets of irreducible matrices

In certain applications of the theory of non-negative matrices, questions pertaining to a *set*, S, of non-negative matrices T become important. For example, in certain dynamic programming problems in the context of mathematical economics, 'optimal' choice of a matrix T, from a set S of available matrices, at any given stage, is relevant; this kind of situation is examined in §3.1. We discuss several such situations from a theoretical viewpoint, together with a brief description of an application. In each case, it will be necessary to impose conditions on the structure of the set S: among these *we shall nearly always assume that any* $T \in S$ *is irreducible.* We shall denote the Perron–Frobenius eigenvalue of T by $r(T)$, and positive corresponding left and right eigenvectors by $\boldsymbol{\nu}'(T)$, $\boldsymbol{\mu}(T)$ respectively.

It is necessary to mention that theory of the kind we discuss here does not appear to be well organized, occurring, as it still does, mainly in the context of various applications. It is therefore probably best to regard the following material as a fairly small selection of topics from the general field. The related topic of inhomogeneous Markov chains (inhomogeneous products of stochastic matrices) is deferred to the next chapter.

3.1 Optimization with a specific set of matrices

In addition to the irreducibility and same dimensions of the component matrices S, we make here the following assumptions:

Assumption 1: S is closed and bounded

Assumption 2: To each (elementwise finite) vector $\boldsymbol{\alpha} \geqslant \mathbf{0}$ and any two matrices $M, N \in S$, there exists a matrix $L \in S$ such

$$L\boldsymbol{\alpha} \geqslant M\boldsymbol{\alpha},\ L\boldsymbol{\alpha} \geqslant N\boldsymbol{\alpha}.$$

The first of these conditions regards each $T \in S$ as a point in Euclidean n^2

space, R_{n^2}, in which S is thus assumed to be compact. The second condition reflects an 'optimal choice' property which is particularly suitable in view of the applicability of the theory of this section; thus whereas Assumption 1, a natural and convenient assumption, will occur elsewhere in the chapter, Assumption 2 will not.

LEMMA 3.1.† Let $\hat{r} = \sup r(T)$. Then $0 < \hat{r} < \infty$ and $\hat{r}$ is attained for some $T \in S$, say T^*, so that

$$\hat{r} = \sup_{T \in S} r(T) = \max_{T \in S} r(T) = r(T^*).$$

Proof. Since an eigenvalue of a matrix is a continuous function of its entries $r(T)$ is a continuous function of $T \in S$. Since S is compact it follows that $r(T)$ is bounded for $T \in S$, and that it attains its supremum for some $T \in S$. #

Now, for a fixed vector $\boldsymbol{\alpha} = \{\alpha_i\} \geqslant \mathbf{0}$, Let us denote by

$$\sup_{T \in S} T\boldsymbol{\alpha}, \text{ the vector of elementwise suprema} \left\{ \sup_{T \in S} \sum_{j=1}^{n} t_{ij} \alpha_j \right\}.$$

LEMMA 3.2. For any fixed i, $\sup \sum_{j=1}^{n} t_{ij} \alpha_j$ is attained for some matrix of the set S, so that we may replace 'sup' by 'max' whenever it occurs. In fact there exists a single matrix for which all the maxima are attained simultaneously, i.e. there exists $\overline{T} \in S$ such that

$$\overline{T}\boldsymbol{\alpha} = \max_{T \in S} T\boldsymbol{\alpha}.$$

Proof. The first part again follows from compactness of S in conjunction with continuity of each elementwise function $\sum_{j=1}^{n} t_{ij} \alpha_j$ in $T \in S$. The second part depends heavily on the 'choice' Assumption 2, and is left as an exercise.‡ #

The following theorem is the analogue, in the present setting, of the Subinvariance Theorem (Theorem 1.6) of Chapter 1. (See also Exercise 2.1)

THEOREM 3.1. If for $\boldsymbol{y} \geqslant \mathbf{0}, \neq \mathbf{0}$ and a fixed positive s

$$\max_{T \in S} T\boldsymbol{y} \leqslant s\boldsymbol{y}, \tag{3.1}$$

then (*a*) $\boldsymbol{y} > \mathbf{0}$; (*b*) $s \geqslant \hat{r}$. Moreover, if $s = \hat{r}$ then equality holds in (3.1), and $\boldsymbol{y} = \boldsymbol{u}(T^*)$, where T^* is any matrix $T \in S$ such that $r(T^*) = \hat{r}$.

Conversely, choosing $\boldsymbol{y}$ as $\boldsymbol{\mu}(T^*)$, (3.1) holds with $s = \hat{r}$ (and hence for $s \geqslant \hat{r}$).

Proof. (3.1) implies that for each $T \in S$

$$T\boldsymbol{y} \leqslant s\boldsymbol{y}. \tag{3.2}$$

† Note that the validity of this lemma does not depend on Assumption 2.
‡ Exercise 3.1.

Hence from the Subinvariance Theorem $s \geqslant r(T)$, and $y > \mathbf{0}$. Thus

$$s \geqslant \sup_{T \in S} r(T) = \hat{r}.$$

Now, suppose $s = \hat{r}$ in (3.1). Then from (3.2) it follows that if T^* is such that $r(T^*) = \hat{r}$,

$$T^* y \leqslant \hat{r} y.$$

From the Subinvariance Theorem again

$$T^* y = \hat{r} y$$

whence $y = \mu(T^*)$, and equality must hold in (3.1).

Conversely, let $y = u(T^*)$ where $r(T^*) = \hat{r}$, and $s = \hat{r}$. Suppose that for this choice of y

$$\max_{T \in S} Ty \leqslant \hat{r} y$$

i.e. selecting $\bar{T}$ as in Lemma 3.2

$$\bar{T} y \leqslant ry, \ \bar{T} \in S.$$

On the other hand

$$T^* y = \hat{r} y.$$

Now, by Assumption 2, we may choose $L \in S$ such that $Ly \geqslant \bar{T}y$, $Ly \geqslant T^* y$,

which implies that $$Ly \geqslant \hat{r} y$$

with strict inequality in at least one position. Multiplying from the left by $v'(L)$ we obtain

$$r(L)\, v'(L)\, y > \hat{r} v'(L)\, y$$

i.e. $$r(L) > \hat{r}$$

which contradicts the definition of $\hat{r}$. #

In the following theorem we examine the asymptotic behaviour of an optimal sequence of $\{\hat{V}_k\}$ defined by

$$\hat{V}_{k+1} = \max_{T \in S} T \hat{V}_k \tag{3.3}$$

where $\hat{V}_0 \geqslant \mathbf{0}, \neq \mathbf{0}$, is arbitrarily chosen. The result proved is in a sense, an analogue in the present setting of the results of Theorem 1.2 and Exercise 1.14 of Chapter 1.

THEOREM 3.2. The n series.

$$\sum_{k=0}^{\infty} z^k \hat{V}_k$$

all have (common) convergence radius $1/\hat{r}$.

Proof. From Theorem 3.1, we may choose $\mu \equiv \mu\,(T^*)$ such that

$$\max_{T \in S} T\mu \leqslant \hat{r}\mu$$

and $\hat{V}_0 \leqslant \mu$ since μ, as an eigenvector, can be chosen to a positive multiple. It follows from the definition of the sequence $\{\hat{V}_k\}$ and induction, that

$$\hat{V}_k \leqslant \hat{r}^k \mu$$

so that each series has convergence radius at least $1/r$.

Now suppose that some series, say the ith:

$$\sum_{k=0}^{\infty} z^k \hat{V}_k\,(i)$$

has convergence radius $R_i > 1/\hat{r}$.

Now, selecting $T^* \in S$ such that $r\,(T^*) = \hat{r}$, define a new vector sequence $\{\bar{V}_k\}$:

$$\bar{V}_k = (T^*)^k\, \bar{V}_0$$

where $$\bar{V}_0 = \hat{V}_0.$$

It follows by induction that

$$\bar{V}_k \leqslant \hat{V}_k,\, k \geqslant 0,$$

and hence for the ith series.

$$\bar{V}_k\,(i) \leqslant \hat{V}_k\,(i).$$

Now, let s satisfy $R_i > 1/s > 1/\hat{r}$. It follows from the nature of R_i that

$$\hat{V}_k\,(i)\,/s^k \rightarrow \mathbf{0}. \tag{3.4}$$

Now, select j such that $\bar{V}_0\,(j) = \hat{V}_0\,(j) > 0$. Then

$$0 \leqslant t^*{}_{ij}^{(k)}\, \bar{V}_0\,(j) \leqslant \bar{V}_k\,(i) \leqslant \hat{V}_k\,(i)$$

for all $k \geqslant 0$, Then (3.4) implies

$$t^*{}_{ij}^{(k)}/s^k \rightarrow 0 \text{ for } 0 < s < \hat{r},$$

which is not possible (see Exercise 1.14). #

Finally we concern ourselves with the existence, uniqueness and nature of solutions $\hat{d}$, $\hat{d} \geqslant \mathbf{0}, \neq \mathbf{0}$ to the equation system

$$s\hat{d} = \boldsymbol{c} + \max_{T \in S} T\hat{d} \tag{3.5}$$

where $\boldsymbol{c} \geqslant \mathbf{0} \neq \mathbf{0}$ is a fixed (but arbitrary) vector, and s is a real number, $s > 0$. The discussion here is a generalization, in the present setting of Theorem 2.1.

For any $T \in S$, we denote by $\boldsymbol{d}\,(T)$ the unique solution

$$(\boldsymbol{d}\,(T) \geqslant \mathbf{0}, \neq \mathbf{0})$$

to the system

$$(sI - T)\,x = c$$

guaranteed by Theorem 2.1 to exist if and only if $s > r(T)$, where s is fixed.

LEMMA 3.3. Let $s > \hat{r}$ and fixed; and $\bar{d} = \sup_{T \in S} d(T)$, in the elementwise sense. Then $\bar{d}$ is attained for some $T \in S$, say $\bar{T}$, so that

$$\bar{d} = d(\bar{T}), \text{ some } \bar{T} \in S.$$

Proof. Let $d_i(T)$ denote the ith element of $d(T)$. Then $d_i(T)$ is a continuous function of $T \in S$ which is compact, so that

$$\sup_{T \in S} d_i(T) \text{ is attained for some } T_i \in S$$

i.e.
$$\sup_{T \in S} d_i(T) = d_i(T_i).$$

Now for any $M,N \in S$, we shall prove that it is possible to find an $L \in S$ such that

$$d(L) \geqslant d(M),\ d(L) \geqslant d(N)$$

from which the result of the Lemma follows.† Suppose then that $M,N \in S$; then put $d^* = \max(d(M), d(N))$ in the elementwise sense. Now

$$sd(M) = c + Md(M) \leqslant c + Md^* \leqslant c + \max_{T \in S} Td^*$$

$$sd(N) = c + Md(N) \leqslant c + Nd^* \leqslant c + \max_{T \in S} Td^* \quad .$$

Hence
$$sd^* \leqslant c + \max_{T \in S} Td^* = c + T_1 d^*$$

for some $T_1 \in S$, by Lemma 3.2. Hence

$$d^* \leqslant c/s + (T_1/s)\,d^* \leqslant (c/s) + (T_1/s)(c/s) + (T_1/s)^2\,d^*$$

and by induction

$$d^* \leqslant s^{-1} \sum_{k=0}^{r-1} (T_1/s)^k\, c + (T_1/s)^r\, d^*.$$

Hence letting $r \to \infty$ (by Lemma B.1 and Exercise 1.14)

since
$$s > \hat{r} \geqslant r(T_1)$$

$$d^* \leqslant s^{-1} \sum_{k=0}^{\infty} (T_1/s)^k\, c = (sI - T_1)^{-1}\, c = d(T_1).$$

Hence
$$d(T_1) \geqslant d(M), d(L), \text{ as required.} \quad \#$$

† Exercise 3.1.

THEOREM 3.3. The equation

$$s\hat{d} = c + \max_{T \in S} T\hat{d}$$

where $c \geqslant 0, \neq 0$, has a solution $\hat{d}$, $\hat{d} \geqslant 0, \neq 0$

(and there is only one such) when $s > \hat{r}$; and has no such solution when

$$0 < s < \hat{r}.$$

Proof. Consider $s > \hat{r}$ and fixed. For arbitrary $T \in S$, $sd(T) = c + Td(T)$

$$\leqslant c + T\bar{d}$$

$$\leqslant c + \max_{T \in S} T\bar{d}$$

so that

$$s\bar{d} \leqslant c + \max_{T \in S} T\bar{d}$$

$$= c + T_1 d$$

for some $T_1 \in S$, i.e.

$$d \leqslant c/s + (T_1/s)\bar{d}. \tag{3.6}$$

Suppose strict inequality in at least one position. Then there is

$$f \geqslant 0, \neq 0,$$

such that

$$\bar{d} + f \leqslant c/s + (T_1/s)\bar{d}. \tag{3.7}$$

Now using (3.6) in the right hand side of (3.7) we obtain

$$\bar{d} + f \leqslant \frac{c}{s} + \left(\frac{T_1}{s}\right)\left(\frac{c}{s} + \frac{T_1\bar{d}}{s}\right)$$

$$= \frac{c}{s} + \frac{T_1}{s}\frac{c}{s} + \left(\frac{T_1}{s}\right)^2 \bar{d}$$

and so, by induction

$$\bar{d} + f \leqslant \sum_{k=0}^{r-1} \left(\frac{T_1}{s}\right)^k \left(\frac{c}{s}\right) + \left(\frac{T_1}{s}\right)^r \bar{d}.$$

Since

$$s > \hat{r} \geqslant r(T_1),$$

it follows as in Lemma 3.3

$$\bar{d} + f \leqslant \left(I - \frac{T_1}{s}\right)^{-1} \left(\frac{c}{s}\right)$$

$$= (sI - T_1)^{-1}\, c = d(T_1)$$

$$\bar{d} + f \leqslant d(T_1)$$

which is a contradiction to the definition of $\bar{d}$. Hence

$$s\bar{d} = c + \max_{T \in S} T\bar{d}. \tag{3.8}$$

Now let d_2, $d_2 \geqslant 0, \neq 0$ be a solution of (3.8) such that $d_2 \neq \bar{d}$. Then

$$sd_2 = c + \max_{T \in S} Td_2$$

i.e. $$sd_2 = c + T_2 d_2 \text{ for some } T_2 \in S. \tag{3.9}$$

Thus $$d_2 = d\,(T_2), \text{ so that}$$

$$d\,(T_2) \leqslant \bar{d}$$

with strict inequality in at least one position, by assumption. Now, since

$$s\bar{d} = c + \bar{T}\bar{d} \tag{3.10}$$

it follows from (3.9) and (3.10) that

$$\begin{aligned} s\,(\bar{d} - d_2) &= \bar{T}\bar{d} - T_2 d_2 \\ &= \bar{T}\bar{d} - \bar{T}d_2 + \bar{T}d_2 - T_2 d_2 \\ &= \bar{T}(\bar{d} - d_2) - k \end{aligned}$$

where $k = -\bar{T}d_2 + T_2 d_2 \geqslant 0$, since $T_2 d_2 = \max_{T \in S} Td_2$.

It follows that

$$\bar{d} - d_2 = -(sI - \bar{T})^{-1} k,$$

from Corollary 1 of Theorem 2.1, and moreover, if k has at least one positive element that

$$d - d_2 < 0$$

which is a contradiction. Thus $k = 0$ and hence $\bar{d} = d_2$ as required.

Now suppose $0 < s < \hat{r}$, and a solution $\hat{d}$ of the required type exists to the equation

$$s\hat{d} = c + \max_{T \in S} T\hat{d}$$

i.e. there is a solution $\hat{d}$, $\hat{d} \geqslant 0, \neq 0$ to $s\hat{d} \geqslant c + T^*\hat{d}$, where $T^* \in S$, $r(T^*) = \hat{r}$. By induction as before

$$\begin{aligned} d &\geqslant s^{-1} \sum_{k=0}^{r-1} (T^*/s)^k\, c + (T^*/s)^r\, \hat{d} \\ &\geqslant (T^*/s)^r\, \hat{d} \end{aligned}$$

and, as we have already seen (in the proof of Theorem 3.2) at least one component of the righ hand side is unbounded as $r \to \infty$, since $0 < s < \hat{r}$. #

COROLLARY. For fixed s, $s > \hat{r}$, the unique solution $\hat{\boldsymbol{d}}$ ($\hat{\boldsymbol{d}} \geq \mathbf{0}, \neq \mathbf{0}$,) of the equation system (3.5) is

$$\bar{\boldsymbol{d}} = \sup_{T \in S} \boldsymbol{d}\,(T).$$

Optimal productivity

In control theory, a linear multistage decision is described by a set

$$S = \{T(u) : u \in U\}$$

of matrices depending on a control parameter u. $T(u)$ is the transformation of the 'state' vector V for the selected value u of the parameter. Thus if V_k is the state vector at stage k of the system, and u_k the kth chosen parameter value, thus

$$V_{k+1} = T(u_k)\, V_k.$$

In the present situation, when S is a set of non-negative irreducible matrices satisfying Assumptions 1 and 2 (2 being a typical dynamic programming assumption), we see that Theorem 3.2 enables us to assert that the *maximal rate of growth* attainable by the system is in fact $\hat{r}$, and could have been attained by a *homogeneous strategy*, i.e. using at each stage the same matrix T^*, where $r\,(T^*) = \hat{r}$. (See Exercise 3.3 for the determination of such a T^* when S is finite.)

Also, consider the dynamic model described at the conclusion of §2.1. The productivity at the $(k + 1)$th stage is *optimized* by choosing a matrix $T = T_k \in S$ such that

$$\boldsymbol{x}_{k+1} = \max_{T \in S} \beta T \boldsymbol{x}_k = \boldsymbol{c}.$$

Then Theorem 3.3 tells us that a stationary regime will exist if $\beta < 1/\hat{r}$, in which case there is only one stationary state vector, given by

$$\boldsymbol{x}_k = \bar{\boldsymbol{d}}, \qquad k = 0, 1, 2, \ldots$$

Moreover $\bar{\boldsymbol{d}} = \bar{d}(\bar{T})$ for some $\bar{T} \in S$, and $\bar{\boldsymbol{d}} \geq \boldsymbol{d}\,(T)$ for $T \in S$, so that this stable point is also the unique stable vector of the 'homogeneous' system

$$\boldsymbol{x}_{k+1} = \beta \bar{T} \boldsymbol{x}_k + c:$$

and the 'production' matrix $\bar{T}$ corresponds to the uniformly greatest (stable) output, out of all $\boldsymbol{d}\,(T)$, $T \in S$,

Bibliography and discussion to §3.1

This section is adapted from the paper of Mandl & Seneta (1969) which treats also the case where the component non-negative matrices are (countably) infinite.

Exercises on §3.1

3.1. Complete the proofs of Lemmas 3.2 and 3.3.

3.2. Show that if S is a set of positive matrices generated by all possible interchanges of corresponding rows selected from a fixed finite set of positive matrices (square and of the same dimension), then S satisfies the Assumptions 1 and 2 of §3.1. (Mandl, 1967)

3.3. Let S be a *finite* collection of irreducible $(n \times n)$ matrices satisfying Assumptions 1 and 2. Define the sequence $\{M_k\}$ $k = 0, 1, 2, \ldots$ where $M_k \in S$, recursively by taking $M_0 \in S$ arbitrary, and letting

$$M_{k+1}\, \mu(M_k) = \max_{M \in S} M\mu\,(M_k),\ k \geqslant 0;$$

in accordance with Lemma 3.2. Show that

$$r(M_k) \leqslant r(M_{k+1})$$

with strict inequality unless we can choose $M_{k+1} = M_k$ {which will eventually occur, since S is a finite set}. Show that if $\hat{k}$ is the first k such that we can choose $M_{k+1} = M_k$, then $r(M_{\hat{k}}) = \hat{r}$. (Use Theorem 3.1.)

(Mandl & Seneta, 1969)

3.2 Inhomogeneous products of non-negative matrices

In the previous section, the question of asymptotic behaviour of an inhomogeneous product of non-negative irreducible matrices was touched on in a rather special setting. This question is of interest in demographic and other contexts, and we deal with it in its own right in this section.

We *assume* that the set S, $S = \{H_k\}$ consists of an ordered sequence of $(n \times n)$ non-negative matrices H_k, $k = 1, 2, \ldots$ satisfying the two conditions:

Assumption A: $T_{p,r} \equiv H_{p+1}\, H_{p+2} \ldots H_{p+r} > 0$ for $p \geqslant 0$ and $r \geqslant r_0$, where $r_0 \geqslant 1$ is some fixed integer independent of p;

Assumption B: $\min^{+}_{i,j}\ h_{ij}\,(k) \geqslant \lambda > 0$

$$\max_{i,j} h_{ij}\,(k) \leqslant \gamma < \infty$$

uniformly for all $k = 1, 2, \ldots$, where $H_k = \{h_{ij}\,(k)\}$, and $\min^{+}$ refers to the minimum among all *positive* entries.

The nature of the conditions may be understood from the particular case where all H_k are identical; Assumption A then asserts primitivity, and B imposes no further restriction.

We shall be concerned under these conditions with the asymptotic behaviour of the product

$$T_{0,k} \equiv H_1 H_2 \ldots H_k \equiv \prod_{r=1}^{k} H_r$$

as $k \to \infty$. (Similar assumptions and methods can be used to study the 'reverse' product $H_k H_{k-1} \ldots H_2 H_1$ as $k \to \infty$; see Exercise 3.6.) To prove the basic theorem we need two preliminary results. The first is not restricted to the present setting either in applicability or scope and expresses a very useful *averaging* (or contraction) property of stochastic matrices P (recall that a stochastic matrix $P = \{p_{ij}\}$, $i, j = 1, \ldots, n$ is a non-negative matrix whose rows sum to unity, i.e. $\sum_j p_{ij} = 1$, all i). The second result is relevant to the present setting under the present assumption only.

LEMMA 3.4. Let P be an $n \times n$ stochastic matrix and w and n-component vector of real elements (not necessarily non-negative), having maximum component M_0 and minimum component m_0; and let M_1 and m_1 be the maximum and minimum components of the vector Pw. Then $M_1 \leqslant M_0$, $m_1 \geqslant m_0$, and

$$M_1 - m_1 \leqslant (1 - 2\epsilon)(M_0 - m_0)$$

where $\epsilon \geqslant 0$ is the minimum entry† of P.

Proof. Let w^* be the vector obtained from w by replacing all components, except one component having value m_0, by M_0; then $w \leqslant w^*$. Each component of the vector Pw^* is then of the form

$$\alpha m_0 + (1 - \alpha) M_0 = M_0 - \alpha (M_0 - m_0);$$

and since $\alpha \geqslant \epsilon$, it follows that each component $\leqslant M_0 - \epsilon(M_0 - m_0)$.

Now, since $w \leqslant w^*$, it follows that

$$M_1 \leqslant M_0 - \epsilon(M_0 - m_0).$$

Applying this result to the vector $-w$ we have

$$-m_1 \leqslant -m_0 - \epsilon(-m_0 + M_0)$$

so that, by addition

$$M_1 - m_1 \leqslant M_0 - m_0 - 2\epsilon(M_0 - m_0) = (1 - 2\epsilon)(M_0 - m_0)$$

as required. #

LEMMA 3.5. Writing $T_{p,r} = \{t_{i,j}^{(p,r)}\}$ under the Assumptions A and B

$$0 < K_1 \leqslant t_{i,j}^{(p,r_0)} \leqslant K_2 < \infty$$

uniformly for all i, j, p, where K_1 and K_2 are constants.

Proof. For all i, j, p

$$\lambda^{r_0} \leqslant t_{i,j}^{(p,r_0)} \leqslant n^{r_0 - 1} \gamma^{r_0}.$$

The left hand inequality follows from the fact that, since

$$t_{i,j}^{(p,r_0)} > 0$$

† And so clearly $\epsilon \leqslant \frac{1}{2}$.

(by Assumption A above), it must contain at least one positive product component (path):

$$t_{ik_1}^{(p,1)}\, t_{k_1k_2}^{(p+1,1)} \ldots t_{k_{r_0-1},j}^{(p+r_0-1,1)}$$
$$\equiv h_{ik_1}(p+1)\, h_{k_1k_2}(p+2) \ldots h_{k_{r_0-1},j}(p+r_0) \geqslant \lambda^{r_0}$$

by Assumption B. The upper bound follows from Assumption B by replacing every entry of the H_k composing T_{p,r_0} by the maximum, γ. #

THEOREM 3.4. (Weak Ergodicity Theorem). Under the Assumptions A and B of this section, as $k \to \infty$, for all i, j, p, s:

$$\frac{t_{i,s}^{(p,k)}}{t_{j,s}^{(p,k)}} \to V_{i,j}^{(p)} > 0$$

where the limit $V_{i,j}^{(p)}$ is independent of s (i.e. the rows of $T_{p,k}$ tend to *proportionality* as $k \to \infty$).

Proof. It is sufficient to prove the proposition for $p = 0$, as we shall do, since the proposition for arbitrary p is no more general.

Now, consider i, j fixed but arbitrary,

$$\frac{t_{i,s}^{(0,k+1)}}{t_{j,s}^{(0,k+1)}} = \frac{\sum_{r=1}^{n} t_{i,r}^{(0,k)}\, h_{rs}(k+1)}{t_{j,s}^{(0,k+1)}}$$

$$= \sum_{r=1}^{n} \frac{t_{i,r}^{(0,k)}}{t_{j,r}^{(0,k)}} \frac{t_{j,r}^{(0,k)}\, h_{rs}(k+1)}{t_{j,s}^{(0,k+1)}}$$

i.e.
$$\frac{t_{i,s}^{(0,k+1)}}{t_{j,s}^{(0,k+1)}} = \sum_{r=1}^{n} d_{s,r}^{(k)} \frac{t_{i,r}^{(0,k)}}{t_{j,r}^{(0,k)}} \qquad (3.11)$$

where
$$d_{s,r}^{(k)} = \frac{t_{j,r}^{(0,k)}\, h_{rs}(k+1)}{t_{j,s}^{(0,k+1)}}$$

noting that for k sufficiently large $t_{i,s}^{(0,k)}$, $t_{j,s}^{(0,k)}$ are positive for all $s = 1, 2, \ldots, n$, by Assumption A. Further, the matrix $\{d_{s,r}^{(k)}\}$ is stochastic, since

$$\sum_r d_{s,r}^{(k)} = \frac{\sum_r t_{j,r}^{(0,k)}\, h_{rs}(k+1)}{t_{j,s}^{(0,k+1)}} = \frac{t_{j,s}^{(0,k+1)}}{t_{j,s}^{(0,k+1)}} = 1.$$

Hence by Lemma 3.4, (for k sufficiently large) applied to (3.11), it follows that if we put

$$M_k = \max_s \frac{t_{i,s}^{(0,k)}}{t_{j,s}^{(0,k)}}, \quad m_k = \min_s \frac{t_{i,s}^{(0,k)}}{t_{j,s}^{(0,k)}}$$

that
$$m_k \leqslant m_{k+1} \leqslant M_{k+1} \leqslant M_k.$$

It follows therefore, by monotonicity and boundedness that

$$m_k \uparrow m^*, M_k \downarrow M^*, \quad \text{where } m^* \leqslant M^*$$

and the assertions of the theorem will follow if we can prove that $m^* = M^*$, or equivalently, that the monotonic decreasing positive sequence $\{M_k - m_k\} \to 0$.

To achieve this, we effectively shall modify the above argument with a changed time scale; thus first we put

$$\overline{T}_{0,k} = T_{0,r_0k},\ k = 1, 2, \ldots$$

putting, analogously,

$$\overline{H}_k = \{\overline{h}_{ij}(k)\} = H_{(k-1)r_0+1} \ldots H_{kr_0} = T_{(k-1)r_0, r_0}$$

so that

$$\overline{T}_{0,k} = \overline{H}_1\overline{H}_2 \ldots \overline{H}_k,$$

$$\overline{T}_{p,r} = \overline{H}_{p+1}\overline{H}_{p+2} \ldots \overline{H}_{p+r}.$$

Moreover condition A implies the appropriate condition:

$(\overline{A})$ $$\overline{H}_r > 0, \text{ all } r = 1, 2, \ldots \text{ (i.e. } \overline{r}_0 = 1)$$

while condition B, which has not been used in the proof to date, implies via Lemma 3.5 that for all i, j, r

$$0 < K_1 \leqslant \overline{h}_{ij}(r) \leqslant K_2 < \infty. \tag{3.12}$$

With the obvious changes to 'bar' notation we have therefore from the previous part of the proof that for all $k \geqslant 1$, $\overline{m}_k \uparrow$ and bounded, $\overline{M}_k \downarrow$ and bounded, and, further from Lemma 3.4

$$\overline{M}_{k+1} - \overline{m}_{k+1} \leqslant (1 - 2\,\overline{\epsilon}_k)(\overline{M}_k - \overline{m}_k)$$

where $\overline{\epsilon}_k$ is the minimal element of the stochastic matrix with general term

$$d_{s,r}^{(k)} = \frac{\overline{t}_{j,r}^{(0,k)}\,\overline{h}_{r,s}(k+1)}{\overline{t}_{j,s}^{(0,k+1)}} = \frac{\overline{t}_{j,r}^{(0,k)}\,\overline{h}_{r,s}(k+1)}{\sum_q \overline{t}_{j,q}^{(0,k)}\,\overline{h}_{q,s}(k+1)}$$

$$= \overline{h}_{r,s}(k+1) \Big/ \sum_q \frac{\overline{t}_{j,q}^{(0,k)}}{\overline{t}_{j,r}^{(0,k)}}\,\overline{h}_{q,s}(k+1). \tag{3.13}$$

We aim at obtaining a uniform bound (independent of k) greater than zero for $\overline{\epsilon}_k$. In view of (3.12) we need only investigate the quantities

$$\frac{\overline{t}_{j,q}^{(0,k)}}{\overline{t}_{j,r}^{(0,k)}}.$$

Now, consider temporarily q and r fixed. Then for each j

$$\frac{\bar{t}_{j,q}^{(0,k+1)}}{\bar{t}_{j,r}^{(0,k+1)}} = \frac{\sum_p \bar{h}_{j,p}(1)\, \bar{t}_{p,q}^{(1,k)}}{\bar{t}_{j,r}^{(0,k+1)}} = \frac{\sum_p \bar{h}_{j,p}(1)\, \bar{t}_{p,r}^{(1,k)}\, \bar{t}_{p,q}^{(1,k)}/\bar{t}_{p,r}^{(1,k)}}{\bar{t}_{j,r}^{(0,k+1)}}$$

$$= \sum_p b_{j,p}^{(k)} \frac{\bar{t}_{p,q}^{(1,k)}}{\bar{t}_{p,r}^{(1,k)}}$$

where

$$b_{j,p}^{(k)} = h_{j,p}^{(1)}\, \bar{t}_{p,r}^{(1,k)}/\bar{t}_{j,r}^{(0,k+1)},$$

so that $\{b_{j,p}^{(k)}\}$ is a stochastic matrix for each k. It therefore follows from Lemma 3.4 that for fixed q, r, and all k

$$\max_p \frac{\bar{t}_{p,q}^{(0,k+1)}}{t_{p,r}^{(0,k+1)}} \leqslant \max_p \frac{\bar{t}_{p,q}^{(0,k)}}{\bar{t}_{p,r}^{(0,k)}} \leqslant \ldots$$

$$\leqslant \max_p \frac{\bar{t}_{p,q}^{(0,1)}}{\bar{t}_{p,r}^{(0,1)}}$$

$$= \max_p \frac{\bar{h}_{p,q}^{(1)}}{\bar{h}_{p,r}^{(1)}}$$

for fixed but arbitrary q and r,

$$\leqslant K_2/K_1$$

by (3.12) which bound is independent of q and r and hence

$$\frac{\bar{t}_{p,q}^{(0,k)}}{\bar{t}_{p,r}^{(0,k)}} \leqslant K_2/K_1$$

for all p, q, r, k.

It thus follows from (3.13) that

$$\bar{\epsilon}_k \geqslant K_1/K_2 K_1^{-1}\, n\, K_2 = n^{-1}\, (K_1/K_2)^2 > 0$$

which is independent of k. It therefore follows that for $k \geqslant 1$

$$\bar{M}_{k+1} - \bar{m}_{k+1} \leqslant (1 - 2\epsilon)\,(\bar{M}_k - \bar{m}_k)$$

where $\epsilon = n^{-1}\, (K_1/K_2)^2$, so that, by iteration, it follows that

$$\bar{M}_k - \bar{m}_k \to 0, \text{ or } k \to \infty.$$

Thus, *returning to the original context*

$$M_{kr_0} - m_{kr_0} \to 0$$

as $k \to \infty$. It thus follows that a subsequence of the convergent sequence $\{M_k - m_k\}$ converges to zero, whence this must also be true for the sequence itself, i.e.

$$M_k - m_k \to 0$$

as required. #

COROLLARY 1. $V_{i,j}^{(0)} = \nu_i/\nu_j,\ i, j = 1, \ldots, n$

where $$\nu = \{\nu_i\} > 0$$

Proof.

$$V_{i,j}^{(0)} = \lim_{k\to\infty} \frac{t_{i,s}^{(0,k)}}{t_{j,s}^{(0,k)}} \lim_{k\to\infty} \frac{t_{i,s}^{(0,k)}/t_{q,s}^{(0,k)}}{t_{j,s}^{(0,k)}/t_{q,s}^{(0,k)}} = \frac{V_{i,q}^{(0)}}{V_{j,q}^{(0)}}$$

for every $q = 1, \ldots, n$. Thus

$$V_{i,j}^{(0)} = \frac{\sum_q V_{i,q}^{(0)}}{\sum_q V_{j,q}^{(0)}} = \frac{\nu_i}{\nu_j}$$

where ν_i may be taken as $\sum_q V_{i,q}^{(0)}$. #

COROLLARY 2.

$$\left| \frac{t_{i,s}^{(0,k)}}{t_{j,s}^{(0,k)}} - V_{i,j}^{(0)} \right| \leqslant K\rho^k$$

where $0 < \rho < 1$ and $0 \leqslant K < \infty$, neither depending on i, j.

Proof. We have that for $k \geqslant 1$

$$M_{kr_0} - m_{kr_0} \leqslant K_{ij}\,(1 - 2\epsilon)^k$$

where $0 \leqslant K_{ij} < 1$ is a constant depending on i, j in general, since i and j were assumed held fixed but arbitrary throughout the main portion of the proof. Thus we can say that for *every* pair i, j

$$M_{kr_0} - m_{kr_0} \leqslant \widetilde{K}\,(1 - 2\epsilon)^k$$

where $\widetilde{K} = \max_{i,j} K_{ij}$. From this, and the fact that the sequence $\{M_k - m_k\}$ is positive and monotonic decreasing, for any pair i, j,

$$M_k - m_k \leqslant \widetilde{K}\,(1 - 2\epsilon)^{k/r_0 - 1}$$

for all $k \geqslant r_0$, and hence by adjusting the constant $\widetilde{K}$,

$$\leqslant K\,\rho^k$$

for all k, where ρ can be taken as

$$(1 - 2\epsilon)^{1/r_0}, \text{ and } \epsilon = n\,(\lambda/n\gamma)^{2r_0}.$$

The assertion of the Corollary follows trivially. The actual estimate of the rate, ρ, of geometric conveyence produced in this proof may be expected to be very conservative in general. #

There is a variant of this theorem worth consideration here,† obtained by

† Some other cases are left to the exercises.

replacing the Assumptions A and B made about the sequence $S = \{H_k\}$ of non-negative matrices by ones which are essentially more restrictive:

Assumption A': H_k has no row of zeros, for each $k = 1, 2, \ldots$
Assumption B': $H_k \to H$ elementwise as $k \to \infty$, where H is a primitive matrix.

LEMMA 3.6. Conditions A' and B' *together* imply conditions A and B.

Proof. See Exercise 3.8.

THEOREM 3.5. (Strong Ergodicity Theorem). Under Assumptions A' and B', for each $i \geq 1$ and $p \geq 0$, as $k \to \infty$

$$\frac{t_{ij}^{(p,k)}}{\sum_{s=1}^{n} t_{i,s}^{(p,k)}} \to v_j, \; j = 1, 2, \ldots, n$$

where $v' \equiv \{v_j\}$ is the positive left Perron–Frobenius eigenvector of H, normed so that $v'\mathbf{1} = 1$. (i.e. the sum-*normed* rows of $T^{(p,k)}$ *all tend to equality* with the sum-normed left eigenvector of H).

Proof. Since the conditions and conclusion are invariant to shift forward by p, we shall only need to prove the theorem in the case $p = 0$.

We shall present a sequence of preliminary steps, followed by the main argument.

1. We first define another sequence of non-negative matrices $\{H_k^{(L)}\}$ as follows, where h_{ij}, $h_{ij}(k)$, $h_{ij}^{(L)}(k)$ denote the general (i, j) elements of the matrices H, H_k, $H_k^{(L)}$, respectively. Let k_0 be such that for $k \geq k_0$, H_k has positive entries in *at least* the same positions as H; that k_0 exists is guaranteed by Assumption B'. Then

$$\begin{aligned} h_{ij}^{(L)}(k) &= \min\{h_{ij}(k), h_{ij}\}, \text{ for all } k \geq k_0 \\ &= h_{ij}(k), \text{ for } 1 \leq k \leq k_0. \end{aligned}$$

The sequence $\{H_k^{(L)}\}$ satisfies conditions B' and A'; moreover for $k \geq k_0$

$$H \geq H_k^{(L)}, \text{ i.e. } H - H_k^{(L)} \geq 0;$$

and
$$H_1^{(L)} H_2^{(L)} \ldots H_k^{(L)} \leq T_{0,k} \equiv H_1 H_2 \ldots H_k$$

since
$$H_k^{(L)} \leq H_k \text{ for all } k \geq 1.$$

2. Further we denote by $\boldsymbol{u}$ the positive right Perron–Frobenius eigenvector of H satisfying $v'u = 1$, and r the Perron–Frobenius eigenvalue. Put also $R = uv', > 0$. Then by Theorem 1.2 of Chapter 1

$$\frac{H^k}{r^k} \to R$$

as $k \to \infty$. Thus we may find a null sequence $\{\delta_k\}$ of numbers $1 \geqslant \delta_k \geqslant 0$ such that

$$(1 - \delta_k) R \leqslant (r^{-1} H)^k \leqslant (1 + \delta_k) R. \tag{3.14}$$

3. Now, in view of the properties of H_k, $H_k^{(L)}$ in relation to H, we may also find null sequences $\{\alpha_k\}$, $\{\beta_k\}$ such that

$$H_k \leqslant H + (\beta_k r)R,\ \beta_k \geqslant 0 \text{ for } k \geqslant 1$$
$$H_k^{(L)} \geqslant H - (\alpha_k r)R,\ \alpha_k \geqslant 0 \text{ for } k \geqslant k_0.$$

4. In conclusion to the preliminaries, we show that if $\{A_k\}$ is a sequence of matrices satisfying $0 \leqslant A_k \leqslant H/r$, $A_k \leqslant \gamma_k R$ for some sequence $\{\gamma_k\}$ with $\gamma_k \geqslant 0$, $k \geqslant 1$, then

$$(r^{-1} H - A_1)(r^{-1} H - A_2) \ldots (r^{-1} H - A_k) \geqslant (r^{-1} H)^k - \sum_{q=1}^{k} \gamma_q R.$$

We proceed by induction: the proposition is obviously true for $k = 1$. Assume it is true for arbitrary $k \geqslant 1$; and multiply from the right by $(r^{-1} H - A_{k+1}) \geqslant 0$. Then

$$\begin{aligned}
&(r^{-1} H - A_1)(r^{-1} H - A_2) \ldots (r^{-1} H - A_{k+1}) \\
&\geqslant (r^{-1} H)^k (r^{-1} H - A_{k+1}) - \left(\sum_{q=1}^{k} \gamma_q R\right)(r^{-1} H) \\
&= (r^{-1} H)^{k+1} - (r^{-1} H)^k A_{k+1} - \sum_{q=1}^{k} \gamma_q R,
\end{aligned}$$

from the nature of R,

$$\geqslant (r^{-1} H)^{k+1} - \gamma_{k+1} H^k - \sum_{q=1}^{k} \gamma_q R$$

using $A_{k+1} \leqslant \gamma_{k+1} R$ and the nature of R

$$= (r^{-1} H)^{k+1} - \sum_{q=1}^{k+1} \gamma_q R$$

as required.

Applying this for $k - m + 1 \geqslant k_0$

$$\begin{aligned}
(r^{-1} H)^m - \sum_{q=k-m+1}^{k} \alpha_q R &\leqslant r^{-m} H_{k-m+1}^{(L)} H_{k-m+2}^{(L)} \ldots H_k^{(L)} \\
&\leqslant r^{-m} H_{k-m+1} H_{k-m+2} \ldots H_k \\
&\leqslant (r^{-1} H + \beta_{k-m+1} R)(r^{-1} H + \beta_{k-m+2} R) \ldots (r^{-1} H + \beta_k R) \\
&= r^{-m} H^m + \left\{\prod_{q=k-m+1}^{k} (1 + \beta_q) - 1\right\} R
\end{aligned}$$

from the nature of R. Thus, invoking (3.14)

$$\{1-\delta_m-\sum_{q=k-m+1}^{k}\alpha_q\}R\leqslant r^{-m}H_{k-m+1}H_{k-m+2}\dots H_k$$

$$\leqslant\left\{\delta_m+\prod_{q=k-m+1}^{k}(1+\beta_q)\right\}R. \tag{3.15}$$

Now consider an arbitrary vector $\boldsymbol{x}\geqslant\boldsymbol{0},\neq\boldsymbol{0}$ and the vector

$$\boldsymbol{w}'\equiv\boldsymbol{x}'H_1H_2\dots H_{k-m}.$$

In virtue of Assumption A$'$, $\boldsymbol{w}\geqslant\boldsymbol{0},\neq\boldsymbol{0}$. On account of (3.15), it follows that

$$r^m\left\{1-\delta_m-\sum_{q=k-m+1}^{k}\alpha_q\right\}\boldsymbol{w}'R\leqslant\boldsymbol{x}'T_{0,k}$$

$$\leqslant r^m\left\{\delta_m+\sum_{q=k-m+1}^{k}(1+\beta_q)\right\}\boldsymbol{w}'R$$

and a similar inequality for $\boldsymbol{x}'T_{0,k}\boldsymbol{u}$ (where, recall, $\boldsymbol{u}$ is the right Perron–Frobenius eigenvector of H), obtained by multiplying throughout by $\boldsymbol{u}$, from the right. Thus combining these two sets of inequalities

$$\frac{\left\{1-\delta_m-\sum_{q=k-m+1}^{k}\alpha_q\right\}\boldsymbol{w}'R}{\left\{\delta_m+\sum_{q=k-m+1}^{k}(1+\beta_q)\right\}\boldsymbol{w}'R\boldsymbol{u}}\leqslant\frac{\boldsymbol{x}'T_{0,k}}{\boldsymbol{x}'T_{0,k}\boldsymbol{u}}$$

$$\leqslant\frac{\left\{\delta_m+\sum_{q=k-m+1}^{k}(1+\beta_q)\right\}\boldsymbol{w}'R}{\left\{1-\delta_m-\sum_{q=k-m+1}^{k}\alpha_q\right\}\boldsymbol{w}'R\boldsymbol{u}}$$

where, from the nature of R,

$$\frac{\boldsymbol{w}'R}{\boldsymbol{w}'R\boldsymbol{u}}=\boldsymbol{v}'$$

independently of $\boldsymbol{w}$, and so of $\boldsymbol{x}$. Now, k and m are arbitrary positive integers, subject only to the constraints $k-m\geqslant k_0-1$, so that, letting first $k\to\infty$, and then $m\to\infty$, it follows that

$$\lim_{k\to\infty}\frac{\boldsymbol{x}'T_{0,k}}{\boldsymbol{x}'T_{0,k}\boldsymbol{u}}=\boldsymbol{v}' \tag{3.16}$$

uniformly with respect to $\boldsymbol{x}\geqslant\boldsymbol{0},\neq\boldsymbol{0}$. Also, clearly

$$\lim_{k\to\infty}\frac{\boldsymbol{x}'T_{0,k}\boldsymbol{1}}{\boldsymbol{x}'T_{0,k}\boldsymbol{u}}=1 \tag{3.17}$$

from which the result of the theorem follows by putting $\boldsymbol{x}'$ as the vector with one in the ith position, zeroes elsewhere, and dividing (3.16) by (3.17). #

COROLLARY.

$$\lim_{k\to\infty} \frac{x' T_{0,k}}{x' T_{0,k} \mathbf{1}} = v'$$

elementwise, converging *uniformly* with respect to $\boldsymbol{x}$, $\boldsymbol{x} \geqslant \mathbf{0}, \neq \mathbf{0}$.

Applications in demography

A simple demographic model for the evolution of the age structure of a human population over a set of 'time points' $k = 0, 1, 2, \ldots$ may be described as follows: if μ_k is an $(n \times 1)$ vector whose components give the numbers in various age groups at time k, then this vector is assumed to change over time according to the recurrence

$$\mu'_{k+1} = \mu'_k H_{k+1}, \quad k = 0, 1, 2, \ldots$$

where H_{k+1} is a known matrix with non-negative entries, depending (in general) on time k, and expressing mortality–fertility conditions at that time. Moreover, as a consequence, the set $S = \{H_k\}$ of matrices has a rather special structure, in that all H_k have the same primitive incidence matrix,† and the (coincident) positive entries of the matrices are bounded uniformly away from zero and infinity; thus certainly Assumptions A and B of this section are satisfied, and hence Theorem 3.4 can be used to make inferences about

$$\begin{aligned} \mu'_k &= \mu'_0 H_1 H_2 \ldots H_k \\ &= \mu'_0 T_{0,k} \end{aligned}$$

where μ_0 ($\geqslant \mathbf{0}, \neq \mathbf{0}$) represents the initial age structure of the population, and $S = \{H_k\}$ the history of mortality–fertility pressures over time. Thus consider two different initial population structures: $\boldsymbol{\alpha} = \{\alpha_i\}$, $\beta = \{\beta_i\}$ subjected to the same 'history' $\{H_k\}$. Then from Theorem 3.4 (dividing numerator and denominator below by $t_{q,s}^{(0,k)}$ for some fixed q) as $k \to \infty$:

$$\frac{\sum_{i=1}^{n} \alpha_i t_{i,s}^{(0,k)}}{\sum_{j=1}^{n} \beta_j t_{j,s}^{(0,k)}} \to \frac{\sum_{i=1}^{n} \alpha_i V_{i,q}^{(0)}}{\sum_{j=1}^{n} \beta_j V_{j,q}^{(0)}} = \frac{\sum_{i=1}^{n} \alpha_i v_i}{\sum_{j=1}^{n} \beta_j v_j}$$

(the last step being a consequence of Corollary 1 of Theorem 3.4) where the limit value is independent of s. This independence of s is called the *weak* ergodicity property, since the μ_k arising from different μ_0 tend to *proportionality* for large k i.e. tend *to represent the same age structure which, however, will still tend to vary with k.*

The property of *strong* ergodicity relates to the situation where the common age structure after a long time k will also tend to remain constant as k

† Of rather special sort: see Exercise 3.5.

increases: this situation obtains if the 'transition matrix' H_k is either independent of k, or tends to become so as k increases: as evidenced by Theorem 3.5 and its Corollary.

Bibliography and discussion to §3.2

A version of Theorem 3.4 was proved, in a demographic setting† by Lopez (1961), and in the demographic literature is generally called the Coale–Lopez theorem. The proof given here is, in most *essential* features, close to that of Lopez, although the statement is of a somewhat more general nature. Both proofs were written under the influence of the approach to the basic ergodic theorem for Markov chains with primitive transition matrix occurring in the book of Kemeny & Snell (1960, pp. 69–70) where the present Lemma 3.4 is given as Theorem 4.1.3. Lemma 3.4 and a kind of dual lemma (Lemma 4.1 of Chapter 4) embody a 'contraction-type property' of stochastic matrices, whose usage (to prove ergodic theorems for both homogeneous and inhomogeneous Markov chains) goes back to Kolmogorov (1931), and, indeed, Markov (1907, 1924) himself. The reader whose interests lie chiefly in demography should consult also the paper of McFarland (1969), who has attempted to argue the validity of the Coale–Lopez theorem from a more (mathematically) elementary standpoint.

Theorem 3.5 is similar in statement and proof, though more general, to a peripheral result given by Joffe & Spitzer (1966, pp. 416–17).

The results of this section may be regarded as attempts to generalize Theorem 1.2 (of Chapter 1) for *powers* of a non-negative matrix to *inhomogeneous* products of non-negative matrices. A great deal of such theory has been developed for the special situation where all H_i are *stochastic,* in the context of inhomogeneous Markov chains, the stochasticity restriction being responsible for substantially better results. We shall take up this situation briefly in the next chapter.

Exercises on §3.2

3.4. Show that if $S = \{H_k\}$ consists of $(n \times n)$ non-negative matrices all of which are irreducible and only a finite number of which have zero entries, then Assumption A is satisfied.

3.5. If $S = \{H_k\}$ consists of $(n \times n)$ non-negative matrices, all of which have the same irreducible incidence matrix (i.e. all of which are irreducible, with positive entries in the same positions) with at least one positive diagonal entry, show that condition A is satisfied with $r_0 = 2(n-1)$.

3.6. Write down assumptions and results analogous to those of this section for the 'backwards' products

$$T_{0,r}^{B} \equiv H_r H_{r-1} \ldots H_1.$$

† See Exercise 3.5 for the mathematical framework.

Hint: Consider the transpose of $T_{0,r}^{B}$.

3.7. Show that the conclusion of Theorem 3.4 is unchanged if condition B is replaced by the condition that

$$0 < \alpha \leqslant \frac{h_{i,j}^{(k)}}{h_{p,q}^{(k)}} \leqslant \beta < \infty$$

(uniformly for all i, j, p, q, k) when $h_{i,j}(k)$ and $h_{p,q}(k)$ are both positive.)

Hint: Use Theorem 3.4 on the sequence $\{\tilde{H}_k\}$ of matrices, where $\tilde{H}_k$ is formed by dividing H_k throughout by one of its positive elements.

3.8. Prove Lemma 3.6.

3.9. Let P_1 be a positive stochastic matrix, and P_2 another with the same left Perron–Frobenius eigenvector $\boldsymbol{\nu}'$ ($\boldsymbol{\nu}' \geqslant \mathbf{0}$, $\boldsymbol{\nu}'\mathbf{1} = 1$); e.g. $P_2 = (1/2)(I + P_1)$. Note that the right Perron–Frobenius eigenvector of an irreducible stochastic matrix is always $\mathbf{1}$, and that the product of stochastic matrices is also stochastic.
Define for $k \geqslant 0$

$$H_{2k+2} = P_1, \; H_{2k+1} = P_2.$$

Now, show that

$$\lim_{k\to\infty} T_{0,k} = \mathbf{1}\boldsymbol{\nu}'$$

even though H_k does not approach a limit matrix (so that *condition* B$'$ for Theorem 3.5 does not hold, but the *conclusion* of the theorem holds, in essence).

3.3 Some other topics

1. *Uniform asymptotic behaviour of powers of primitive matrices*
One of the most useful and important results of the Perron–Frobenius theory in regard to the asymptotic behaviour of non-negative matrices, is expressed by Theorem 1.2 which implies that for a *primitive* matrix T, as $k \to \infty$

$$\{T^k/r^k - \boldsymbol{\mu}\boldsymbol{\nu}'\} \to 0$$

where $\boldsymbol{\mu}$, $\boldsymbol{\nu}'$ are positive left and right eigenvectors of T corresponding to the Perron–Frobenius eigenvalue r, and normed so that $\boldsymbol{\nu}'\boldsymbol{\mu} = 1$ (before, we denoted these vectors by $\mathbf{w}$, $\boldsymbol{\nu}'$ respectively).

In some applications, such as multitype branching process stability theory,† where an infinite *set* S of $n \times n$ primitive matrices is being considered, it is important to know that the approach to the zero matrix in (3.18) is *uniform* with respect to $T \in S$. We shall prove that this is so under the general assumption that
($\mathscr{A}$) The set S of primitive matrices T is closed and bounded (i.e. compact, if the $T \in S$ are regarded as points in Euclidean n^2-space, R_{n^2}).

† See Quine (1972).

This type of assumption has already been untilized in §3.1 of this chapter, and we shall again make use of the obvious notation $r(T)$, $\mu(T)$, $\nu'(T)$ introduced there, with the additional assumption $\nu'(T)\,\mu(T) = 1$. We shall also denote by $\lambda_2(T)$ an eigenvalue of T with largest modulus after $r(T)$. In fact, we shall find it easier to work with the family of matrices M derived from the corresponding $T \in S$ by $M = T/r(T)$, so that $\mu(M) = \mu(T)$, $\nu'(M) = \nu'(T)$, $r(M) = 1$, $\lambda_2(M) = \lambda_2(T)/r(T)$, so that $|\lambda_2(M)| < 1$.

Before proceeding, we point out that a discussion such as that which we now undertake must of necessity pertain substantially to the *entire spectrum,* not just the pleasant dominant Perron–Frobenius structure, so that it will in due course be necessary to deal to some extent with matrices which have complex-valued entries. (The reader who may not be acquainted with some of the methods used in the sequel, such as unitary transformations, should read Appendix C.)

LEMMA 3.7. The family of primitive matrices $M(T)$, where $T \in S$, is bounded, and

$$|\lambda_2(M)| < \delta < 1 \tag{3.19}$$

uniformly for all $T \in S$, for some fixed δ.

Proof. It is easy to show, following the pattern of Lemma 3.1 (of §3.1) that $r(T)$ is bounded away from zero for $T \in S$, so that the uniform boundedness of the Ms follows from that of the $T \in S$.

Suppose now (3.19) is false; there exists a sequence $\{T_i\} \subset S$ such that

$$|\lambda_2(M_i)| \to 1$$

as $i \to \infty$, where $M_i = T_i/r(T_i)$. Now, since S is bounded, there is a *convergent* subsequence $\{T_{i_j}\}$, $j = 1, 2, \ldots$ of $\{T_i\}$, $i = 1, 2, \ldots$; and since S is closed

$$\lim_{j\to\infty} T_{i_j} = \widetilde{T} \in S$$

(the limit can be regarded as elementwise). Moreover, since $|\lambda_2(M)|$ is a continuous function of $T \in S$, we have, as a consequence

$$|\lambda_2(\widetilde{M})| = 1, \text{ where } \widetilde{M} = \widetilde{T}/r(\widetilde{T}),$$

which is a contradiciton, since $|\lambda_2(\widetilde{M})| < 1$ by definition. #

In the following theorem the proof is for $n \geqslant 2$; the case $n = 1$ is trivial.

THEOREM 3.6. Under assumption ($\mathscr{A}$) on the S, for $T \in S$

$$\{M^k(T) - \mu(T)\nu'(T)\} \to 0$$

elementwise as $k \to \infty$, uniformly for $T \in S$.

Proof. Consider $M \equiv M(T)$ for $T \in S$; then by Lemma C.1 (Schur's Theorem) of Appendix C, there exists a unitary matrix V such that

$$V^{-1}MV = U = \begin{bmatrix} 1 & s' \\ \mathbf{0} & N \end{bmatrix}$$

where U is an $n \times n$ upper triangular matrix, with 1 in the (1, 1) position, so that N is also upper triangular. Moreover all the diagonal elements (eigenvalues) of N do not exceed the constant $\delta < 1$, specified by the preceding Lemma 3.7 in modulus. This last implies immediately that $N^k \to 0$ as $k \to \infty$†, which we shall, however, now prove directly, so as not to rely on external material.

Clearly

$$(V^{-1}MV)^k = V^{-1}M^kV = \begin{bmatrix} 1 & s'\sum_{i=0}^{k-1} N^i \\ \mathbf{0} & N^k \end{bmatrix}. \tag{3.20}$$

Since we know that the limit of the left hand matrix exists,

$$\widetilde{N} = \lim_{k \to \infty} N^k$$

exists, and, further since the diagonal elements of N^k are $0(\delta^k)$ as $k \to \infty$, $\widetilde{N}$ has its diagonal elements (as well as subdiagonal) zero. Moreover

$$\widetilde{N} = N\widetilde{N} = N^2\widetilde{N} = \ldots = N^r\widetilde{N}$$

for arbitrary r, so that, letting $r \to \infty$,

$$\widetilde{N} = \widetilde{N}^2, = \widetilde{N}^3 = \ldots = \widetilde{N}^{n-1}, = 0$$

from the form of $\widetilde{N}$. Thus invoking also Lemma B.1 of the Appendix, and letting $k \to \infty$ in (3.20)

$$V^{-1}\mu\nu'V = \begin{bmatrix} 1 & s'(I-N)^{-1} \\ \mathbf{0} & 0 \end{bmatrix}. \tag{3.21}$$

Subtracting (3.21) from (3.20):

$$V^{-1}[M^k - \mu\nu']V = \begin{bmatrix} 0 & -s'\sum_{i=k}^{\infty} N^i \\ \mathbf{0} & N^k \end{bmatrix}$$

and applying the Euclidean norm, noting V is unitary

$$\|M^k - \mu\nu'\| = \left\| \begin{bmatrix} 0 & -s'\sum_{i=k}^{\infty} N^i \\ \mathbf{0} & N^k \end{bmatrix} \right\|,$$

† By the well-known result of Oldenburger (1940).

i.e.

$$\|M^k - \mu v'\| \leqslant \|s' \sum_{i=k}^{\infty} N^i\| + \|N^k\|,$$

the last following as an easily checked property of the Euclidean norm. Now, all the matrices and vectors in the last expression actually depend on the value of $T \in S$ being considered although the dependence has been notationally omitted. To obtain the result of the theorem, which is tantamount to $\|M^k - \mu v'\| \to 0$ *uniformly* for $T \in S$, we shall prove that, as $k \to \infty$, the right hand side is dominated by a quantity approaching zero independently of $T \in S$.

We write

$$N = D + F$$

where D is $(n-1) \times (n-1)$ diagonal, and F is upper triangular, with zero elements on its diagonal, of the same dimensions. Thus $F^{n-1} = 0$, and

$$\|D\| \equiv |\lambda_2(M)| < \delta < 1,$$

where δ is independent of $T \in S$. Moreover, by multiplying out $(D+F)^k$, and noting that for any diagonal matrices D_i, $(D_1F)(D_2F)\ldots(D_{n-1}F) = 0$,

$$\|(D+F)^k\| \leqslant \sum_{i=0}^{n-2} \binom{k}{i} \|D\|^{k-i} \|F\|^i$$

taking $k \geqslant n-2$.

Now, the set $M(T)$, $T \in S$, is bounded by Lemma 3.7 and moreover

$$\|M\| = \|V^{-1}MV\| = \left\| \begin{bmatrix} 1 & s' \\ \mathbf{0} & (D+F) \end{bmatrix} \right\|$$

so that it follows that there are finite bounds K_1, K_2 *independent* of any particular $T \in S$ such that $\|F\| < K_1$, $\|s'\| < K_2$. Thus

$$\|(D+F)^k\| \leqslant \sum_{i=0}^{n-2} \binom{k}{i} \delta^{k-i} K_1^i;$$
$$\leqslant \alpha(n-1) \binom{k}{n-2} \delta^{k-(n-2)}$$

where $\alpha = \max(1, K_1^{n-2})$, for $k \geqslant 2(n-2)$, i.e.

$$\|N^k\| \leqslant \beta k^n \delta^k \tag{3.22}$$

where $\beta = \alpha(n-1)\,\delta^{-(n-2)}/(n-2)!$

and so, for $k \geqslant 2(n-2)$

$$\|s' \sum_{i=k}^{\infty} N^i\| \leqslant \|s'\| \sum_{i=k}^{\infty} \|N^i\|$$

$$\leqslant \beta K_2 \sum_{i=k}^{\infty} i^n \delta^i. \tag{3.23}$$

From (3.22) and (3.23) it follows that as $k \to \infty$

$$\|N^k\| \text{ and } \|s' \sum_{i=k}^{\infty} N^i\| \to 0$$

uniformly with respect to $T \in S$ as required, since the dominating quantity of neither of these norms depends on specific $T \in S$. #

2. *A convexity property of non-negative matrices*

The set S of $(n \times n)$ matrices which we consider in this subsection is that generated by a single matrix $T(\theta) = \{t_{ij}(\theta)\}$ whose *positive* entries depend on a parameter θ, $\theta \in I$: $\theta_1 < \theta < \theta_2$ for some fixed θ_1, θ_2, and remain positive as θ varies (and whose zero entries do not depend on θ).

We shall assume for simplicity that $T(\theta)$, $\theta \in I$ has its incidence matrix *irreducible* (and denote by $r(\theta)$ the Perron–Frobenius eigenvalue of $T(\theta)$), although the irreducibility restriction is easily removed.†

For the necessary notions of convexity and superconvexity occurring in the sequel, see Appendix D.

THEOREM 3.7. If every positive entry $t_{ij}(\theta)$ of $T(\theta)$ is a superconvex function on I (i.e. $\log t_{ij}(\theta)$ is convex on I), then $r(\theta)$ is also (i.e. $\log r(\theta)$ is convex on I).

Proof. If $d \geq 1$ is the period of the matrix $T(\theta)$, $\theta \in I$, let

$$T_1(\theta) = T^d(\theta).$$

Then $T_1(\theta)$ is either primitive ($d = 1$); or contains a primitive 'diagonal' matrix corresponding to a cyclic class. In either case write this primitive matrix as $T_2(\theta)$; its Perron–Frobenius eigenvalue $r_2(\theta) \equiv r^d(\theta)$ – see Chapter 1. Since $T_2(\theta)$ is primitive there exists an n_0 such that

$$T_3(\theta) = T_2^{n_0}(\theta) > 0,$$

and its Perron–Frobenius eigenvalue is

$$r_3(\theta) = r^{n_0 d}(\theta).$$

We shall prove $r_3(\theta)$ is superconvex in I, which is tantamount to the superconvexity of $r(\theta)$.

We first note that since each entry of $T_3(\theta)$ is a sum of products of the positive entries of $T(\theta)$, it is also superconvex in $\theta \in I$.‡

Put

$$\begin{aligned} f_k(\theta) &= \{\text{trace } T_3^k(\theta)\}^{1/k} \\ &= \{\text{ sum of the eigenvalues of } T_3^k(\theta)\}^{1/k} \\ &= r_3(\theta)\{1 + o(\delta^k)\}^{1/k} \end{aligned}$$

† See Exercise 3.10.

‡ Lemma D.3 of Appendix D.

where $0 < \delta < 1$, by the Perron–Frobenius Theory,

$$\rightarrow r_3(\theta)$$

as $k \rightarrow \infty$, where $f_k(\theta)$ is superconvex.† Thus $r_3(\theta)$, a positive limit of superconvex functions on I, is itself superconvex on I, as required. #

COROLLARY. The function $r(\theta)$ is convex in I. [Any superconvex function is convex.]

Bibliography and discussion to §3.3

Theorem 3.6 follows in statement and proof the more general paper of Buchanan & Parlett (1966). Theorem 3.7 was proved by Kingman (1961) under the assumption of positivity of $T(\theta)$, $\theta \in I$.

Exercises on §3.3

(These pertain to subsection 2 of §3.3).

3.10. Show that Theorem 3.2 remains valid without the assumption of irreducibility of the common incidence matrix of each $T(\theta)$, $\theta \in I$, so long as this matrix has *positive* dominant eigenvalue.

3.11. Show that $r(\theta)$ is superconvex if

$$t_{ij}(\theta) = \int_{-\infty}^{\infty} e^{\theta x}\, dF_{ij}(x)$$

where $F_{ij}(x)$ is increasing and such that the integral converges in I for all i, j.

Hint: Use Hölder's inequality for Stieltjes integrals. (Miller, 1961)

† See Lemma D.3 of Appendix D.

4 Markov chains and finite stochastic matrices

Certain aspects of the theory of non-negative matrices are particularly important in connection with that class of simple stochastic processes known as Markov chains, and we feel that it is necessary to give a sketch of such theory, firstly, because the theory of finite Markov chains in part provides a useful illustration of the more widely applicable theory developed hitherto; and, secondly, because some of the theory of countable Markov chains, once developed, can be used as a starting point, as regards ideas, towards an analytical theory of infinite non-negative matrices (as we shall eventually do) which can then be developed without reference to probability notions.

In this chapter, after the introductory concepts, we shall confine ourselves to finite Markov chains, which is virtually tantamount to a study from a certain viewpoint of finite stochastic matrices. In the next chapter we shall pass to the study of countable Markov chains, which is thus tantamount to a study of stochastic matrices with countable index set, which of course will subsume the finite index set case. Thus this chapter in effect concludes an examination of finite non-negative matrices, and the next initiates our study of the countable case.

We are aware that the general reader may not be acquainted with the simple probabilistic concepts used to initiate the notions of these two chapters. Nevertheless, since much of the content of this chapter and the next is merely a study of the behaviour of stochastic matrices, we would encourage him to persist if he is interested in this last, skipping the probabilistic passages. The next chapter, in particular, is almost free of probabilistic notions.

4.1 Markov chains

Informally, Markov chains (MCs) serve as theoretical models for describing a 'system' which can be in various 'states', the fixed set of possible states being countable (i.e. finite, or denumerably infinite). The system 'jumps' at unit time intervals from one state to another, and the probabilistic law according to which jumps occur is

'If the system is in the ith state at time $k-1$, the next jump will take it to the jth state with probability $p_{ij}(k)$'.

The set of transition probabilities $p_{ij}(k)$ is prescribed for all i, j, k and determines the probabilistic behaviour of the system, once it is known how it starts off 'at time 0'.

A more formal description is as follows. We are given a countable set $\mathscr{S} = \{s_1, s_2 \ldots\}$ or, sometimes, more conveniently $\{s_0, s_1, s_2, \ldots\}$ which is known as the state space, and a sequence of random variables $\{X_k\}$, $k = 0, 1, 2, \ldots$ taking values in $\mathscr{S}$, and having the following *probability property:* if $x_0, x_1, \ldots, x_{k+1}$ are elements of $\mathscr{S}$, then

$$P(X_{k+1} = x_{k+1} | X_k = x_k, X_{k-1} = x_{k-1}, \ldots, X_0 = x_0)$$
$$= P(X_{k+1} = x_{k+1} | X_k = x_k)$$

if $$P(X_k = x_k, \ldots, X_0 = x_0) > 0$$

(if $P(B) = 0$, $P(A|B)$ is undefined).

This property which expresses, roughly, that future probabilistic evolution of the process is determined once the *immediate past* is known, is the Markov property, and the stochastic process $\{X_k\}$ possessing it is called *a Markov chain.*

Moreover, we call the probability

$$P(X_{k+1} = s_j | X_k = s_i)$$

the transition probability from state s_i to state s_j, and write it succinctly as

$$p_{ij}(k+1),\ s_i,\ s_j \in \mathscr{S},\ k = 0, 1, 2, \ldots$$

Now consider

$$P[X_0 = s_{i_0}, X_1 = s_{i_1}, \ldots, X_k = s_{i_k}].$$

Either *this is positive,* in which case, by repeated use of the Markov property and conditional probabilities it is in fact

$$P[X_k = s_{i_k} | X_{k-1} = s_{i_{k-1}}] \ldots P[X_1 = s_{i_1} | X_0 = s_{i_0}]\ P[X_0 = s_{i_0}]$$
$$= p_{i_{k-1}, i_k}(k)\, p_{i_{k-2}, i_{k-1}}(k-1) \ldots p_{i_0, i_1}(1)\, \Pi_{i_0}$$

where $\Pi_{i_0} = P[X_0 = s_{i_0}]$

or it is zero, in which case for some $0 \leqslant r \leqslant k$ (and we take such minimal r)

$$P[X_0 = s_{i_0}, X_1 = s_{i_1}, \ldots, X_r = s_{i_r}] = 0.$$

Considering the cases $r = 0$ and $r > 0$ separately, we see (repeating the above argument), that it is *nevertheless* true that

$$P[X_0 = s_{i_0}, X_1 = s_{i_1}, \ldots, X_k = s_{i_k}] = \Pi_{i_0}\, p_{i_0, i_1}(1) \ldots p_{i_{k-1}, i_k}(k)$$

since the product of the first $r+1$ elements on the right is zero. Thus we see that the probability structure of any finite sequence of outcomes is *completely defined* be a knowledge of the *non-negative quantities*

$$p_{ij}(k);\ s_i,\ s_j \in \mathscr{S}$$

$$\Pi_i;\ s_i \in \mathscr{S}.$$

The set $\{\Pi_i\}$ of probabilities is called the *initial probability distribution* of the chain. We consider these quantities as specified, and denote the row vector of the initial distribution by $\boldsymbol{\Pi}_0'$.

Now, for fixed $k = 1, 2, \ldots$ the matrix

$$P_k = \{p_{ij}(k)\},\ s_i, s_j \in \mathscr{S}$$

is called the *transition matrix* of the MC at time k. It is clearly a square matrix with non-negative elements, and will be doubly infinite if $\mathscr{S}$ is denumerably infinite.

Moreover, its row sums (understood in the limiting sense in the denumerably infinite case) are unity, for

$$\sum_{j\in\mathscr{S}} p_{ij}(k) = \sum_{j\in\mathscr{S}} P[X_k = s_j | X_{k-1} = s_i]$$

$$= P[X_k \in \mathscr{S} | X_{k-1} = s_i]$$

by the addition of probabilities of disjoint sets;

$$= 1.$$

Thus the matrix P_k is *stochastic*.

DEFINITION 4.1. If $P_1 = P_2 = \ldots = P_k = \ldots$ the Markov chain is said to have stationary transition probabilities or is said to be *homogeneous.* Otherwise it is *non-homogeneous.*

In the homogeneous case we shall refer to the common transition matrix as *the* transition matrix, and denote it by P.

Let us denote by Π_k' the row vector of the probability distribution of X_k; then it is easily seen from the expression for a single finite sequence of outcomes in terms of transition and initial probabilities that

$$\Pi_k' = \Pi_0' P_1 \ldots P_k$$

by summing (possibly in the limiting sense) over all sample paths for any fixed state at time k. In keeping with the notation of §3.2, we might now adopt the notation

$$T_{p,r} = P_{p+1} P_{p+2} \ldots P_{p+r}$$

and write

$$\Pi_k' = \Pi_0' T_{0,k}.$$

[We digress for a moment to stress that, even in the case of infinite transition matrices, the above products are well defined by the natural extension of the rule of matrix multiplication, and are themselves stochastic. For: let

$$P_\alpha = \{p_{ij}(\alpha)\} \text{ and } P_\beta = \{p_{ij}(\beta)\}$$

be two infinite stochastic matrices defined on the index set $\{1, 2, \ldots\}$. Define their product $P_\alpha P_\beta$ as the matrix with i, j entry given by the (non-negative) number:

$$\sum_{k=1}^{\infty} p_{ik}(\alpha) p_{kj}(\beta).$$

This sum converges, since the summands are non-negative, and

$$\sum_{k=1}^{\infty} p_{ik}(\alpha) p_{kj}(\beta) \leqslant \sum_{k=1}^{\infty} p_{ik}(\alpha) \leqslant 1$$

since probabilities always take on values between 0 and 1. Further the ith row sum of the new matrix is

$$\sum_{j=1}^{\infty} \sum_{k=1}^{\infty} p_{ik}(\alpha) p_{kj}(\beta) = \sum_{k=1}^{\infty} p_{ik}(\alpha) \left(\sum_{j=1}^{\infty} p_{kj}(\beta) \right)$$

$$= \sum_{k=1}^{\infty} p_{ik}(\alpha) = 1$$

by stochasticity of both P_α and P_β. (The interchange of summations is justified by the non-negativity of the summands.)]

It is also easily seen that for $k > p$

$$\Pi'_k = \Pi'_p \, T_{p,k-p}.$$

We are now in a position to see why the theory of homogeneous chains is substantially simpler than that of non-homogeneous ones: for then

$$T_{p,k} = P^k$$

so we have only to deal with powers of the common transition matrix P, and further, the probabilistic evolution *is homogeneous in reference to any initial time point p.*

In the remaining section of this chapter we assume that we are dealing with finite ($n \times n$) *matrices as before,* so that the index set is $\{1, 2, \ldots, n\}$ as before (or perhaps, more conveniently, $\{0, 1, \ldots, n-1\}$).

Examples

(1) *Bernoulli scheme.* Consider a sequence of independent trials in each of which a certain event has fixed probability, p, of occurring (this outcome being called a 'success') and therefore a probability $q = 1 - p$ of not occurring (this outcome being called a 'failure'). We can in the usual way equate success

with the number 1 and failure with the number 0; then $\mathscr{S} = \{0, 1\}$, and the transition matrix at any time k is

$$P = \begin{bmatrix} q & p \\ q & p \end{bmatrix}$$

so that we have here a homogeneous 2-state Markov chain. Notice that here the rows of the transition matrix are identical, which must in fact be so for any 'Markov chain' where the random variables $\{X_k\}$ are independent.

(2) *Random walk between two barriers.* A particle may be at any of the points $0, 1, 2, 3, \ldots, s (s \geqslant 1)$ on the x-axis. If it reaches point 0 it remains there with probability a and is reflected with probability $1 - a$ to state 1; if it reaches point s it remains there with probability b and is reflected to point $s - 1$ with probability $1 - b$. If at any instant the particle is at position i, $1 \leqslant i \leqslant s - 1$, then at the next time instant it will be at position $i + 1$ with probability p, or at $i - 1$ with probability $q = 1 - p$.

It is again easy to see that we have here a homogeneous Markov chain on the finite state set $\mathscr{S} = \{0, 1, 2, \ldots, s\}$ with transition matrix

$$P = \begin{bmatrix} a & 1-a & 0 & 0 \ldots 0 & 0 & 0 & 0 \\ q & 0 & p & 0 \ldots 0 & 0 & 0 & 0 \\ 0 & q & 0 & p \ldots 0 & 0 & 0 & 0 \\ & & & \vdots & & & & \\ 0 & 0 & 0 & 0 \ldots q & 0 & p & 0 \\ 0 & 0 & 0 & 0 \ldots 0 & 0 & 1-b & 0 \end{bmatrix} ; p + q = 1, 0 < p < 1.$$

If $a = 0$, 0 is a *reflecting barrier,* if $a = 1$ it is an *absorbing barrier,* otherwise i.e. if $0 < a < 1$ it is an *elastic barrier;* and similarly for state s.

(3) *Random walk unrestricted to the right.* The situation is as above, except that there is no 'barrier' on the right i.e. $\mathscr{S} = \{0, 1, 2, 3, \ldots\}$ is denumerably infinite, and so is the transition matrix P.

(4) *Recurrent event.* Consider a 'recurrent event', described as follows. A system has a variable lifetime, whose length (measured in discrete units) has probability distribution $\{f_i\}$, $i = 1, 2, \ldots$ When the system reaches age $i \geqslant 1$, it either continues to age, or 'dies' and starts afresh from age 0. The movement of the system if its age is $i - 1$ units, $i \geqslant 2$ is thus to i, with (conditional) probability $(1 - f_1 - \ldots - f_i)/(1 - f_1 - \ldots - f_{i-1})$ or to age 0, with probability $f_i/(1 - f_1 - \ldots - f_{i-1})$. At age $i = 0$, it either reaches age 1 with probability $1 - f_1$, or dies with probability f_1.

We have here a homogeneous Markov chain on the state set $\mathscr{S} = \{0, 1, 2, \ldots\}$ describing the movement of the age of the system. The transition matrix is then the denumerably infinite one:

$$\begin{bmatrix} f_1 & 1-f_1 & 0 & 0 & 0\ldots \\ \dfrac{f_2}{1-f_1} & 0 & \dfrac{1-f_1-f_2}{1-f_1} & 0 & 0\ldots \\ \dfrac{f_3}{1-f_1-f_2} & 0 & 0 & \dfrac{1-f_1-f_2-f_3}{1-f_1-f_2} & 0\ldots \\ & & \vdots & & \end{bmatrix}$$

It is customary to specify only that $\sum_{i=1}^{\infty} f_i \leqslant 1$, thus allowing for the possibility of an infinite lifetime.

(5) *Pólya Urn scheme.* Imagine we have a white and b black balls in an urn. Let $a + b = N$. We draw a ball at random and before drawing the next ball we replace the one drawn, adding also s balls of the same colour.

Let us say that after r drawings the system is in state i, $i = 0, 1, 2, \ldots$ if i is the number of white balls obtained in the r drawings. Suppose we are in state i ($\leqslant r$) after drawing number r. Thus $r - i$ black balls have been drawn to date, and the number of white balls in the urn is $a + is$, and the number of black is $b + (r - i)s$. Then at the next drawing we have movement to state $i + 1$ with probability.

$$p_{i,i+1}(r+1) = \frac{a+is}{N+rs}$$

and to state i with probability

$$p_{i,i}(r+1) = \frac{b+(r-i)s}{N+rs} = 1 - p_{i,i+1}(r+1).$$

Thus we have here a *non-homogeneous* Markov chain (if $s > 0$) with transition matrix P_k at 'time' $k \equiv r + 1 \geqslant 1$ specified by

$$\begin{aligned} p_{ij}(k) &= \frac{a+is}{N+(k-1)s}, \; j = i+1 \\ &= \frac{b+(k-1-i)s}{N+(k-1)s}, \; j = i \\ &= 0 \text{ otherwise,} \end{aligned}$$

where $\mathscr{S} = \{0, 1, 2, \ldots\}$.

N.B. This example is given here because it is a good illustration of a non-homogeneous chain; the non-homogeneity clearly occurring because of the addition of s balls of colour like the one drawn at each stage. Nevertheless, the reader should be careful to note that this example does not fit into the framework in which we have chosen to work in this chapter, since the matrix P_k is really *rectangular,* viz. $k \times (k + 1)$ in this case, a situation which can occur with non-homogeneous chains, but which we omit from further theoretical consideration. Extension in both directions to make each P_k doubly infinite corresponding to the index set $\{0, 1, 2, \ldots\}$ is not necessarily a good idea, since matrix dimensions are equalized at the cost of zero rows (beyond the $(k - 1)$th) thus destroying stochasticity.

4.2 Finite homogeneous Markov chains

Within this section we are in the framework of the bulk of the matrix theory developed hitherto.

It is customary in Markov chain theory to classify states and chains of various kinds. In this respect we shall remain totally consistent with the classification of Chapter 1.

Thus a chain will be said to be *irreducible,* and, further, *primitive* or *cyclic* (*imprimitive*) according to whether its transition matrix P is of this sort. Further, states of the set

$$\mathscr{S} = \{s_1, s_2, \ldots, s_n\}$$

(or $\{s_0, s_1, \ldots, s_{n-1}\}$) will be said to be *periodic, essential and inessential,* to *lead* one to another, to *communicate,* to form *essential and inessential classes* etc. according to the properties of the corresponding indices of the index set $\{1, 2, \ldots, n\}$ of the transition matrix.

In fact, as has been mentioned earlier, this terminology was introduced in Chapter 1 in accordance with Markov chain terminology. The reader examining the terminology in the present framework should now see the logic behind it.

Irreducible MCs

Suppose we consider an irreducible MC $\{X_k\}$ with (irreducible) transition matrix P. Then putting as usual $\mathbf{1}$ for the vector with unity in each position,

$$P\mathbf{1} = \mathbf{1}$$

by stochasticity of P; so that 1 is an eigenvalue and $\mathbf{1}$ a corresponding eigenvector. Now, since all row sums of P are equal and the Perron–Frobenius eigenvalue lies between the largest and the smallest, 1 is the Perron–Frobenius eigenvalue of P, and $\mathbf{1}$ may be taken as the corresponding right Perron–Frobenius eigenvector. Let ν', normed so that $\nu'\mathbf{1}$, be the corresponding positive left eigenvector. Then we have that

$$\nu' P = \nu', \tag{4.1}$$

where ν is the column vector of a probability distribution.

DEFINITION 4.2. Any initial probability distribution Π_0 is said to be *stationary,* if

$$\Pi_0 = \Pi_k, \ k = 1, 2, \ldots;$$

and a Markov chain with such an initial distribution is itself said to be stationary.

THEOREM 4.1. An irreducible MC has a unique stationary distribution given by the solution $\boldsymbol{\nu}$ of $\boldsymbol{\nu}'P = \boldsymbol{\nu}'$, $\boldsymbol{\nu}'\mathbf{1} = 1$.

Proof. Since

$$\Pi'_{k+1} = \Pi'_k P, \ k = 0, 1, 2, \ldots$$

it is easy to see by (4.1) that such $\boldsymbol{\nu}$ is a stationary distribution. Conversely, if Π_0 is a stationary distribution

$$\Pi'_0 = \Pi'_0 P, \ \Pi_0 \geqslant \mathbf{0}, \ \Pi'_0 \mathbf{1} = 1$$

so that by uniqueness of the left Perron–Frobenius eigenvector of P, $\Pi_0 = \boldsymbol{\nu}$. #

THEOREM 4.2. (Ergodic Theorem for primitive MCs). As $k \to \infty$, for a primitive MC

$$P^k \to \mathbf{1}\boldsymbol{\nu}'$$

elementwise where $\boldsymbol{\nu}$ is the unique stationary distribution of the MC; and the rate of approach to the limit is geometric.

Proof. In view of Theorem 4.1, and preceding remarks, this is just a restatement of Theorem 1.2 of Chapter 1 in the present more restricted framework.#

This theorem is extremely important in MC theory for it says that for a primitive MC at least, the probability distribution of X_k, viz. $\Pi'_0 P^k \to \boldsymbol{\nu}'$, which is *independent* of Π_0, and the rate of approach is very fast. Thus, after a relatively short time, past history becomes irrelevant, and the chain approaches a stationary regime.†

We see, in view of the Perron–Frobenius theory that the analytical (rather than probabilistic) reasons for this are (i) $r = 1$, (ii) $w = 1$.

We leave here the theory of irreducible chains, which can be further developed without difficulty via the results of Chapter 1.

Reducible chains with some inessential states

We know from Lemma 1.1 of Chapter 1 that there is always *at least one* essential class associated with a finite MC. Let us assume P is in canonical form as in Chapter 1, §1.2, and that Q is the submatrix of P associated with transitions between the inessential states. We recall also that in P^k we have Q^k in the position of Q in P.

† See also Theorem 4.6 and its following notes.

THEOREM 4.3. $Q^k \to 0$ elementwise as $k \to \infty$, geometrically fast.

Proof. [We could here invoke the classical result of Oldenburger (1940); however we have tried to avoid this result in the present text, since we have nowhere proved it, and so we shall prove Theorem 4.3 directly. In actual fact. Theorem 4.3 can be used to some extent to replace the need of Oldenburger's result for reducible non-negative matrices.]

Any inessential state leads to an essential state.† Let the totality of essential indices of the chain be denoted by E, and of the inessential matrices by I.

We have then that

$$1 - \sum_{j \in I} p_{ij}^{(k)} = \sum_{j \in E} p_{ij}^{(k)} > 0$$

for some k, for any fixed $i \in I$, so that

$$\sum_{j \in I} p_{ij}^{(k)} < 1.$$

Now $\sum_{j \in I} p_{ij}^{(k)}$ is *non-increasing with k,* for

$$\sum_{j \in I} p_{ij}^{(k+1)} = \sum_{j \in I} \sum_{r \in I} p_{ir}^{(k)} p_{rj}$$

$$\leqslant \sum_{r \in I} p_{ir}^{(k)}.$$

Hence for $k \geqslant k_0(\mathrm{i})$ and some k_0 (i)

$$\sum_{j \in I} p_{ij}^{(k)} < \theta(i) < 1$$

and since the number of indices in I is finite, we can say that for $k \geqslant k_0$, and $\theta < 1$, where k_0 and θ are independent of i,

$$\sum_{j \in I} p_{ij}^{(k)} < \theta < 1, \textit{all } i \in I.$$

Therefore

$$\sum_{j \in I} p_{ij}^{(mk+k)} = \sum_{r \in I} p_{ir}^{(mk)} \sum_{j \in I} p_{rj}^{(k)}$$

$$\leqslant \theta \sum_{r \in I} p_{ir}^{(mk)}$$

for fixed $k \geqslant k_0$, and each $m \geqslant 0$ and $i \in I$. Hence

$$\sum_{j \in I} p_{ij}^{(k(m+1))} \leqslant \theta \sum_{r \in I} p_{ir}^{(mk)} \leqslant \theta^{m+1} \to 0 \text{ as } m \to \infty.$$

Hence a subsequence of

$$\sum_{j \in I} p_{ij}^{(k)}$$

approaches zero; but since this quantity is itself positive and monotone non-increasing with m, it has a limit also, and must have the same limit as the subsequence.

† See Exercise 4.11.

Hence $$Q^k \mathbf{1} \to \mathbf{0} \text{ as } k \to \infty.$$

and hence $$Q^k \to 0. \quad \#$$

Now, if the process $\{X_k\}$ passes to an essential state, it will stay forever after in the essential class which contains it. Thus the process cannot ever return to or pass to the set I from the essential states, E. Hence if $\Pi_0(I)$ is that subvector of the initial distribution vector which corresponds to the inessential states we have from the above theorem that

$$\begin{aligned} P[X_k \subset I] &= \Pi_0'(I) Q^k \mathbf{1} \\ &\to 0 \end{aligned}$$

as $k \to \infty$, which can be seen to imply, in view of the above discussion, that the process $\{X_k\}$ leaves the set I of states in a finite time with probability 1 i.e. the process is eventually 'absorbed', with probability 1, into the set E of essential states.

Denote now by E_ρ a specific essential class, $(\bigcup_\rho E_\rho = E)$, and let $x_{i\rho}$ be the probability that the process is eventually absorbed into E_ρ, so that

$$\sum_\rho x_{i\rho} = 1,$$

having started at state $i \in I$. Let $x_{ip}^{(1)}$ be the probability of absorption after precisely one step, i.e.

$$x_{i\rho}^{(1)} = \sum_{j \in E_\rho} p_{ij}, \; i \in I,$$

and let x_ρ and $x_\rho^{(1)}$ denote the column vectors of these quantities over $i \in I$.

THEOREM 4.4.

$$x_\rho = [I - Q]^{-1} x_\rho^{(1)}$$

Proof. First of all we note that since $Q^k \to 0$ as $k \to \infty$ by Theorem 4.3 $[I - Q]^{-1}$ exists by Lemma B.1 of Appendix B (and $= \sum_{k=0}^{\infty} Q^k$ elementwise).

Now let $x_{i\rho}^{(k)}$ by the probability of absorption by time k into E_ρ from $i \in I$. Then the elementary theorems of probability, plus the Markov property enable us to write

$$x_{i\rho}^{(k)} = x_{i\rho}^{(1)} + \sum_{r \in I} p_{ir} x_{r\rho}^{(k-1)} \quad \text{(Backward Equation)},$$

$$x_{ip}^{(k)} = x_{ip}^{(k-1)} + \sum_{r \in I} p_{ir}^{(k-1)} x_{r\rho}^{(1)} \quad \text{(Forward Equation)}.$$

The Forward Equation tells us that

$$(1 \geqslant)\, x_{i\rho}^{(k)} \geqslant x_{i\rho}^{(k-1)}$$

so that $\lim_{k\to\infty} x_{i\rho}^{(k)}$ exists, and it is plausible to interpret this (and it can be rigorously justified) as $x_{i\rho}$.

If we now take limits in the Backward Equation as $k \to \infty$

$$x_{i\rho} = x_{i\rho}^{(1)} + \sum_{r\in I} p_{ir}\, x_{r\rho},$$

an equation whose validity is intuitively plausible. Rewriting this in matrix terms,

$$\boldsymbol{x}_\rho = \boldsymbol{x}_\rho^{(1)} + Q\boldsymbol{x}_\rho$$

$$(I - Q)\boldsymbol{x}_\rho = \boldsymbol{x}_\rho^{(1)}$$

from which the statement of the theorem follows. #

The matrix $[I - Q]^{-1}$ plays a vital role in the theory of finite absorbing chains (as does its counterpart in the theory of transient infinite chains to be considered in the next chapter) and it is sometimes called the *fundamental matrix* of absorbing chains. We give one more instance of its use.

Let Z_{ij} be the number of visits to state $j \in I$ starting from $i \in I$. ($Z_{ii} \geq 1$). Then

$$Z_i = \sum_{j\in I} Z_{ij}, \quad i \in I$$

is the time to absorption of the chain starting from $i \in I$. Let $m_{ij} = \mathcal{E}(Z_{ij})$ and $m_i = \mathcal{E}(Z_i)$ be the expected values of Z_{ij} and Z_i respectively, and $M = \{m_{ij}\}_{i,j\in I}$, and $\boldsymbol{m} = \{m_i\}$.

THEOREM 4.5.

$$M = (I - Q)^{-1}$$

$$\boldsymbol{m} = M\boldsymbol{1} = (I - Q)^{-1}\boldsymbol{1}.$$

Proof. Recall that $Q^0 = I$ by definition. Let $Y_{ij}^{(k)} = 1$ if $X_k = j$, $Y_{ij}^{(k)} = 0$ if $X_k \neq j$, the process $\{X_k\}$ having started at $i \in I$. Then

$$\mathcal{E}\,(Y_{ij}^{(k)}) = p_{ij}^{(k)} \,.\, 1 + (1 - p_{ij}^{(k)}) \,.\, 0$$

$$= p_{ij}^{(k)}, \; k \geq 0.$$

Moreover

$$Z_{ij} = \sum_{k=0}^{\infty} Y_{ij}^{(k)}$$

the sum on the right being effectively finite for any realization of the process, since absorption occurs in finite time. By positivity

$$m_{ij} = \mathcal{E}(Z_{ij}) = \sum_{k=0}^{\infty} \mathcal{E}(Y_{ij}^{(k)}) = \sum_{k=0}^{\infty} p_{ij}^{(k)},$$

$i, j \in I$. Thus

$$M = \sum_{k=0}^{\infty} Q^k \quad \text{elementwise}$$

$$= (I - Q)^{-1} \quad \text{(Lemma B.1 of Appendix B)}$$

and since

$$Z_i = \sum_{j \in I} Z_{ij},$$

it follows that

$$m_i = \sum_{j \in I} m_{ij}. \quad \#$$

Finally, in connection with the fundamental matrix, the reader may wish to note that, in spite of the elegant matrix forms of Theorems 4.4 and 4.5, it may still be easier to solve the corresponding linear equations for the desired quantities in actual problems. These are

$$x_{ip} = x_{ip}^{(1)} + \sum_{r \in I} p_{ir}\, x_{rp}, \; i \in I. \text{ (Theorem 4.4)}$$

$$m_i = 1 + \sum_{r \in I} p_{ir}\, m_r, \; i \in I. \text{ (Theorem 4.5)}$$

and we shall do so in the following example.

Example: Random walk between two absorbing barriers (see §4.1)
Here there are two essential classes E_0, E_s consisting of one state each (the absorbing barriers). The inessential states are $I = \{1, 2, \ldots, s-1\}$ where we assume $s > 1$, and the matrix Q is given by

$$Q = \begin{bmatrix} 0 & p & 0 \ldots 0 & 0 \\ q & 0 & p \ldots 0 & 0 \\ & & \vdots & \\ 0 & 0 & 0 \ldots 0 & q \end{bmatrix}$$

with $x_{i0}^{(1)} = \delta_{i1}\, q$, $x_{is}^{(1)} = \delta_{i,s-1}\, p$, $i \in I$, δ_{ij} being the Kronecker delta.

(*a*) *Probability of eventual absorption into* E_s. We have that

$$x_{is} = x_{is}^{(1)} + \sum_{r \in I} p_{ir}\, x_{rs}, \; i = 1, 2, \ldots, s-1,$$

i.e.

$$\begin{aligned} x_{1s} &= \phantom{q x_{1s}} \; p\, x_{2s} \\ x_{2s} &= q\, x_{1s} + p\, x_{3s} \\ &\;\vdots \\ x_{s-2,s} &= q\, x_{s-3,s} + p\, x_{s-1,s} \\ x_{s-1,s} &= p + q\, x_{s-2,s}. \end{aligned}$$

Write for convenience $x_i \equiv x_{is}$. Then if we define $x_0 = 0, x_s = 1$, the above equations can be written in unified form as

$$\begin{cases} x_i = q\,x_{i-1} + p\,x_{i+1}, \; i = 1, 2, \ldots, s-1 \\ x_0 = 0, x_s = 1 \end{cases}$$

and it is now a matter of solving this difference equation under the stated boundary assumptions.

The general solution is of the form

$$\begin{aligned} x_i &= A\,z_1^i + B\,z_2^i \qquad \text{if } z_1 \neq z_2 \\ &= (A + Bi)z^i \qquad \text{if } z_1 = z_2 = z \end{aligned}$$

where z_1 and z_2 are the solutions of the characteristic equation

$$p\,z^2 - z + q = 0$$

viz., $z_1 = 1, z_2 = q/p$
bearing in mind that $1 - 4pq = (p - q)^2$.
Hence:

(i) if $q \neq p$, we get, using boundary conditions to fix A and B

$$x_i = \{1 - (q/p)^i\}/\{1 - (q/p)^s\}, \quad i \in I.$$

(ii) if $q = p = \dfrac{1}{2}$

$$x_i = i/s, \quad i \in I.$$

(*b*) *Mean time to absorption.* We have that

$$m_i = 1 + \sum_{r \in I} p_{ir}\, m_r, \quad i \in I$$

i.e.

$$\begin{aligned} m_1 &= 1 \qquad\quad + pm_2 \\ m_2 &= 1 + qm_1 + pm_2 \\ &\vdots \\ m_{s-1} &= 1 + qm_{s-2}. \end{aligned}$$

Hence we can write in general

$$\begin{cases} m_i = 1 + qm_{i-1} + pm_{i+1}, \quad i = 1, 2, \ldots, s-1 \\ m_0 = 0, m_s = 0. \end{cases}$$

We have here to deal with an inhomogeneous difference equation, the homogeneous part of which is as before, so that the general solution to it is as before plus a particular solution to the inhomogeneous equation. It can be checked that

(i) $q \neq p$; $i/(q-p)$ is a particular solution, and that taking into account boundary conditions

$$m_i = i/(q-p) - \{s/(q-p)\}\{1-(q/p)^i\}/\{1-(q/p)^s\}, \quad i \in I.$$

(ii) $q = p = \frac{1}{2}$; $-i^2$ is a particular solution and hence

$$m_i = i(s-i), \quad i \in I.$$

To conclude the discussion of absorbing chains we give (i) a slightly generalized ergodic theorem referring to the long-term behaviour of a Markov chain with possibly some inessential states; (ii) a pseudo-ergodic theorem referring to the inessential states above. In this connection it is useful to define at this stage a regular† stochastic matrix, and hence a regular Markov chain as one with a regular transition matrix.

DEFINITION 4.3. An $n \times n$ stochastic matrix is said to be *regular* if its essential indices form a single essential class, which is aperiodic.

THEOREM 4.6. Let P be the transition matrix of a regular MC, in canonical form, and $\boldsymbol{\nu}_1'$ the stationary distribution corresponding to the primitive submatrix P_1 of P corresponding to the essential states. Let $\boldsymbol{\nu}' = (\boldsymbol{\nu}_1', \mathbf{0}')$ be an $1 \times n$ vector. Then as $k \to \infty$

$$P^k \to \mathbf{1}\boldsymbol{\nu}'$$

elementwise, where $\boldsymbol{\nu}'$ is the unique stationary distribution corresponding to the matrix P, the approach to the limit being geometrically fast.

Proof. Apart from the limiting behaviour of $p_{ij}^{(k)}$, $i \in I$, $j \in E$, this theorem is a trivial consequence of foregoing theory, in this and the preceding section.

If we write

$$P^k = \begin{bmatrix} P_1^k & 0 \\ R_k & Q^k \end{bmatrix}$$

it is easily checked (by induction, say) that putting $R_1 = R$

$$R_{k+1} = \sum_{i=0}^{k} Q^i R P_1^{k-i} = \sum_{i=0}^{k} Q^{k-i} R P_1^i$$

so that we need to examine this matrix as $k \to \infty$.

Put $M = P_1 - \mathbf{1}\boldsymbol{\nu}_1'$;

then $$M^i = P_1^i - \mathbf{1}\boldsymbol{\nu}_1'.$$

Now from Theorem 4.2 we know that each element of M^i is dominated by $K_1\rho_1^i$, for some $K_1 > 0$, $0 < \rho_1 < 1$, independent of i, for every i.

† Our definition of 'regular' differs from that of Kemeny & Snell (1960).

Moreover

$$R_{k+1} = \sum_{i=0}^{k} Q^{k-i} R\, 1 \nu_1' + \sum_{i=0}^{k} Q^{k-i} R\, M^i$$

and we also know from Theorem 4.3 that each element of Q^i is dominated by $K_2 \rho_2^i$ for some $K_2 > 0$, $0 < \rho_2 < 1$ independent of i, for every i. Hence each component of the right hand sum matrix is dominated by

$$K_3 \sum_{i=0}^{k} \rho_2^{k-i} \rho_1^i$$

for some $K_3 > 0$, and hence $\to 0$ as $k \to \infty$.

Hence, as $k \to \infty$

$$\lim_{k\to\infty} R_{k+1} = \sum_{i=0}^{\infty} Q^i R\, 1\, \nu_1' = (I - Q)^{-1} R\, 1\, \nu_1'$$
$$= (I - Q)^{-1} (I - Q)\, 1\, \nu_1'$$
$$= 1\, \nu_1'$$

as required. #

Both Theorems 4.2 and 4.6 express conditions under which the probability distribution: $P[X_k = j], j = 1, 2, \ldots, n$ approaches a limit distribution $\nu = \{\nu_j\}$ as $k \to \infty$, *independent of the initial distribution* $\Pi = \{\Pi_j\}$ *of* $\{X_k\}$. Thus

$$\lim_{k\to\infty} P[X_k = j] = \nu_j,\ j = 1, 2, \ldots, n$$

where

$$\nu' P = \nu',\ \nu' 1 = 1.$$

The tendency to a limiting distribution independent of the initial distribution expresses a tendency to equilibrium regardless of initial state; and is called the *ergodic property*.†

The following theorem is analogous to Theorem 4.2 in that when attention is focused on behaviour within the set I of *inessential states*, under similar structural conditions on the set I, then a totally analogous result obtains.

THEOREM 4.7. Let Q, the submatrix of P corresponding to transitions between the inessential states of the MC corresponding to P, be primitive, and let there be a positive probability of $\{X_k\}$ beginning in some $i \in I$. Then for $j \in I$ as $k \to \infty$

$$P[X_k = j | X_k \in I] \to \nu_j^{(2)} / \sum_{j\in I} \nu_j^{(2)}$$

where $\nu^{(2)} = \{\nu_j^{(2)}\}$ is a positive vector independent of the initial distribution, and is, indeed, the left Perron–Frobenius eigenvector of Q.

† See Exercises 4.9 and 4.10.

Proof. Let us note that if Π_0 is that part of the initial probability vector restricted to the initial states then

$$P[X_k \in I] = \Pi_0' Q^k \mathbf{1}, > 0$$

since by primitivity $Q^k > 0$ for k large enough, and $\Pi_0 \neq \mathbf{0}$. Moreover the vector of the quantities

$$P[X_k = j | X_k \in I], j \in I$$

is given by

$$\Pi_0' Q^k / \Pi_0' Q^k \mathbf{1}.$$

The limiting behaviour follows on letting $k \to \infty$ from Theorem 1.2 of Chapter 1, the contribution of the right Perron–Frobenius eigenvector dropping out between numerator and denominator. #

Bibliography and discussion to §§4.1–4.2

There exists an enormous literature on finite homogeneous Markov chain theory; the concept of a Markov chain is generally attributed to A. A. Markov (1907), although some recognition is also accorded to H. Poincaré in this connection. We list here only the *books* which have been associated with the significant development of this subject, and which may thus be regarded as milestones in its development, referring the reader to these for further earlier references: Markov (1924), Hostinsky (1931), von Mises (1931), Fréchet (1938), Bernstein (1946), Romanovsky (1949), Kemeny & Snell (1960). [The reader should notice that these references are not quite chronological, as several of the books cited appeared in more than one edition, the latest edition being generally mentioned here.] An informative sketch of the early history of the subject has been given by W. Doeblin (1938), and we adapt it freely here for the reader's benefit, in the next two paragraphs.

After the first world war the topic of homogeneous Markov chains was taken up by Urban, Lévy, Hadamard, Hostinsky, Romanovsky, von Mises, Fréchet and Kolmogorov. Markov himself had considered the case where the entries of the finite transition matrix $P = \{p_{ij}\}$ were all positive, and showed that in this case all the $p_{ij}^{(k)}$ tend to a positive limit independent of the initial state, s_i, a result rediscovered by Lévy, Hadamard, and Hostinsky. In the general case ($p_{ij} \geqslant 0$) Romanovsky (under certain restrictive hypotheses) and Fréchet, in noting the problem of the calculation of the $p_{ij}^{(k)}$ was essentially an algebraic one, showed that the $p_{ij}^{(k)}$ are asymptotically periodic, Fréchet then distinguishing three situations: the positively regular case, where $p_{ij}^{(k)} \to p_j > 0$, all i, j; the regular case, where $p_{ij}^{(k)} \to p_j \geqslant 0$ all i, j; the non-oscillating case where $p_{ij}^{(k)} \to p_j^i$, for all i, j; and also the general singular case. Fréchet linked the discussion of these cases to the roots of the characteristic equation of the matrix P. Hostinsky, von Mises, and Fréchet found necessary and sufficient conditions for positive regularity.† Finally, Hadamard (1928) gave, in the special case pertaining to card shuffling, the reason for the asymptotic periodicity which enters in the singular case, by using non-algebraic reasoning.

On the other hand the matrix $P = \{p_{ij}\}$ is a matrix of non-negative elements; and these matrices were studied extensively before the first world war by Perron and, particularly, by Frobenius. The remarkable results of Frobenius which enable one to analyse immediately the singular case, were not utilized until somewhat later in chain theory. The first person to do so was probably von Mises (1931), who, in his treatise, deduced a number of important theorems for the singular case from the results of Frobenius. The schools of Fréchet and of Hostinsky remained unaware of this component of von Mises' works, and ignorant also of the third memoir of Frobenius on non-negative matrices. In 1936 Romanovsky, certainly equally unaware of the same work of von Mises, deduced, also from the theorems of Frobenius, by a quite laborious method, theorems more precise than those of von Mises. Finally, Kolmogorov gave in 1936 a complete study of Markov chains with a countable number of states, which is applicable therefore to finite chains.

The present development of the theory of finite homogeneous Markov chains is no more than an introduction to the subject, as the reader will now realise; it deals, further, only with ergodicity problems, whereas there are many problems more probabilistic in nature, such as the Central Limit Theorem, which have not been touched on, because of the nature of the present book. Our approach is of course, basically from the point of view (really a consequence) of the Perron–Frobenius theory, into which elements of the Kolmogorov approach have been blended. The reader interested in a somewhat similar, early, development, would do well to consult Doeblin's (1938) paper; and a sequel by Sarymsakov (1945).

Some further discussion pertaining *specifically* to the case of countable, rather than finite state space (or, correspondingly, index set) will be found in the next chapter.

Exercises on §4.2

(All these exercises refer to homogeneous Markov chains.)

4.1. Let P be an irreducible stochastic matrix, with period $d = 3$. Consider the asymptotic behaviour, as $k \to \infty$, of P^{3k}, P^{3k+1}, P^{3k+2} respectively, in relation to the unique stationary distribution corresponding to P. Extend to arbitrary period d.

Hint: Adapt Theorem 1.4 of Chapter 1.

† See Exercise 4.10 as regards the regular case; Doeblin omits these references.

4.2. Find the unique stationary distribution vector ν for a random walk between two reflecting barriers, assuming s is odd. (See Example (2) of §4.1.) Apply the results of Exercise 4.1, to write down $\lim_{k\to\infty} P^{2k}$ and $\lim_{k\to\infty} P^{2k+1}$ in terms of the elements of ν.

4.3. Use either the technique of Appendix B, or induction, to find P^k for arbitrary k, where (stochastic) P is given by

$$\text{(i)} \begin{pmatrix} p_1 & q_1 \\ p_2 & q_2 \end{pmatrix}; \quad \text{(ii)} \begin{pmatrix} 1 & 0 & 0 \\ 0 & 1 & 0 \\ \frac{1}{3} & \frac{1}{3} & \frac{1}{3} \end{pmatrix}; \quad \text{(iii)} \begin{pmatrix} 0 & 1 & 0 \\ q & 0 & p \\ 0 & 1 & 0 \end{pmatrix}.$$

4.4. Consider two urns A and B, each of which contains m balls, such that the total number of balls, $2m$, consists of equal numbers of black and white members. A ball is removed simultaneously from each urn and put into the other at times $k = 1, 2, \ldots$

Explain why the number of white balls in urn A before each transfer forms a Markov chain, and find its transition probabilities.

Give intuitive reasons why it might be expected that the limiting/stationary distribution $\{\nu_i\}$ of the number of white balls in urn A is given by the hyper-geometric probabilities

$$\nu_i = \frac{\binom{m}{m-i}\binom{m}{i}}{\binom{2m}{m}}, \quad i = 0, 1, 2, \ldots, m$$

and check that this is so.

4.5. Let P be a finite doubly stochastic matrix (i.e. a stochastic matrix all of whose column sums are also unity).

(i) Show that the states of the Markov chain corresponding to P are all essential.

(ii) If P is irreducible and aperiodic, find $\lim P^m$ as $m \to \infty$

4.6. A Markov chain $\{X_k\}$, $k = 0, 1, 2, \ldots$ is defined on the states $0, 1, 2, \ldots, 2N$, its transition probabilities being given by

$$p_{j,i} = \binom{2N}{i}\left(\frac{j}{2N}\right)^i \left(1 - \frac{j}{2N}\right)^{2N-i},$$

$j, i = 0, 1, \ldots, 2N$. Investigate the nature of the states.

Show that for $m \geq 0$, $j = 0, 1, 2, \ldots, 2N$,

$$\mathcal{E}[X_{m+1} | X_m = j] = j,$$

and that consequently

$$\mathcal{E}[X_{m+1} | X_0 = j] = j.$$

Hence, or otherwise, deduce the probabilities of eventual absorption into the state 0 from the other states. (Malécot, 1944)

PHILLIPS MEMORIAL
LIBRARY
PROVIDENCE COLLEGE

4.7. A Markov chain is defined on the integers 0, 1, 2, . . ., a, its transition probabilities being specified by

$$p_{i,i+1} = \frac{1}{2}\left(\frac{a-i-1}{a-i}\right),$$

$$p_{i,i-1} = \frac{1}{2}\left(\frac{a-i+1}{a-i}\right),$$

$i = 1, 2, \ldots, a-1$, with states 0 and a being absorbing.

Find the mean time to absorption, m_i, starting from $i = 1, 2, \ldots, a-1$.

Hints: (1) The state $a - 1$ is reflecting. (2) Use the substitution $z_i = (a-i)m_i$.

4.8. Let L be an $(n \times n)$ matrix with zero elements on the diagonal and above, and U an $(n \times n)$ matrix with zero elements below the diagonal. Suppose that $P_1 = L + U$ is stochastic.

(i) Show that $L^n = 0$, and hence (with the help of probabilistic reasoning, or otherwise), that the matrix $P_2 = (I - L)^{-1}\ U$ is stochastic.

(ii) Show by example that even if P_1 is irreducible and aperiodic, P_2 may be reducible.

4.9. Theorem 4.6 may be regarded as asserting that a sufficient condition for ergodicity is the regularity of the transition matrix P. Show that regularity is in fact a necessary condition also.

4.10. Use the definition of regularity and the result of the preceding exercise to show that a necessary and sufficient condition on the matrix P for ergodicity is that there is only one eigenvalue of modulus unity (counting any repeated eigenvalues as distinct). (Kaucky, 1930; Konečný, 1931)

4.11. Show that any inessential state leads to an essential state.

Hint: Use a contradiction argument, as in the proof of Lemma 1.1 of Chapter 1.

4.3 Finite inhomogeneous Markov chains

In this section, as already foreshadowed in §4.1 of this chapter, we shall adopt the notation of §3.2 of Chapter 3 except that we shall use $P_k = \{p_{ij}(k)\}$ instead of $H_k = \{h_{ij}(k)\}$, $i, j = 1, \ldots, n$, to emphasize the stochasticity of P_k, and we shall be concerned with the asymptotic behaviour of the product

$$T_{0,k} \equiv P_1\, P_2 \ldots P_k \equiv \prod_{r=1}^{k} P_r$$

as $k \to \infty$.

Naturally, both Theorems 3.4 and 3.5 are applicable here, and it is natural to begin by examining their implications in the present context, where, we note, each $T_{p,r}$ is stochastic also.

Under the conditions of the first of these, as $k \to \infty$, for all i, j, p, s

$$\frac{t_{i,s}^{(p,k)}}{t_{j,s}^{(p,k)}} \to V_{ij}^{(p)} > 0$$

where the limit is independent of s. Put for sufficiently large k,

$$\frac{t_{i,s}^{(p,k)}}{t_{j,s}^{(p,k)}} = V_{ij}^{(p)} + \epsilon(i, j, s, p, k).$$

Thus, using the stochasticity of $T_{p,k}$

$$1 = \sum_{s=1}^{n} t_{i,s}^{(p,k)} = V_{ij}^{(p)} + \sum_{s=1}^{n} t_{j,s}^{(p,k)}\ \epsilon(i, j, s, p, k)$$

where also $0 < t_{j,s}^{(p,k)} \leqslant 1$. Letting $k \to \infty$, it follows that the additional assumption has led to:

$$1 = V_{i,j}^{(p)},\ \text{all } i, j, p.$$

so that the rows tend not only to proportionality, but indeed equality, although their nature still depends on k in general. Indeed we may write

$$\frac{t_{i,s}^{(p,k)}}{t_{j,s}^{(p,k)}} - 1 = \frac{t_{i,s}^{(p,k)} - t_{j,s}^{(p,k)}}{t_{j,s}^{(p,k)}} \to 0$$

as $k \to \infty$, which on account of the present boundedness of $t_{j,s}^{(p,k)}$ implies

$$t_{i,s}^{(p,k)} - t_{j,s}^{(p,k)} \to 0 \tag{4.2}$$

as $k \to \infty$, for each i, j, p, s. This conclusion is a weaker one than that preceding it, and so we may expect to obtain it under weaker assumptions (although in the present stochastic context) than given in Theorem 3.4. In fact, since this kind of assertion does not involve a ratio, the condition A imposed in the former context, to ensure positivity of denominator, *inter alia,* may be expected to be subject to weakening.

Under the conditions of Theorem 3.5, in addition to the present stochasticity assumption, we have, simply, that

$$t_{i,j}^{(p,k)} \to v_j,\ j = 1, 2, \ldots, n \tag{4.3}$$

where $\boldsymbol{v}' = \{v_j\}$ is the unique invariant distribution of the limit primitive matrix P.

(4.2) and (4.3) are manifestations of weak and strong ergodicity respectively in the MC sense.

DEFINITION 4.4. We shall say that *weak ergodicity* obtains for the MC (i.e. sequence of stochastic matrices P_i) if

$$t_{i,s}^{(p,k)} - t_{j,s}^{(p,k)} \to 0$$

as $k \to \infty$ for each i, j, s, p. (Note that it is sufficient to consider $i \neq j$.)

This definition does not imply that the $t_{i,s}^{(p,k)}$ themselves tend to a limit as $k \to \infty$, merely that the rows tend to equality (= 'independence of initial distribution') but are still in general dependent on k.

DEFINITION 4.5. If weak ergodicity obtains, and the $t_{i,s}^{(p,k)}$ themselves tend to a limit for all i, s, p as $k \to \infty$, then we say *strong ergodicity* obtains.

Hence strong ergodicity requires the elementwise existence of $T_{p,k}$ as $k \to \infty$ for each p, in addition to weak ergodicity.

A stochastic matrix with identical rows is sometimes called *stable.* Note that if P is stable, $P^2 = P$, and so $P^k = P$.

Thus we may, in a consistent way, speak of weak ergodicity as tendency to stability.

We shall first give a discussion of theory pertaining to weak ergodicity, then pass briefly to strong ergodicity.

Weak ergodicity

A great deal of the theory of weak ergodicity of inhomogeneous Markov chains may be developed with the aid of the following fundamental result, which is a kind of dual result to Lemma 3.4 of Chapter 3 in expressing contractive properties of stochastic matrices.

LEMMA 4.1. Let $P = \{p_{ij}\}$ be an $n \times n$ stochastic matrix and $\delta = \{\delta_i\}$ an n-component vector of real elements satisfying $\boldsymbol{\delta} \neq \mathbf{0}$, $\boldsymbol{\delta}'\mathbf{1} = 0$. Let $\Delta_0 = \Sigma|\delta_i|$ and $\Delta_1 = \Sigma|\delta_i^*|$ where $\boldsymbol{\delta}^* = \{\delta_i^*\}$ is defined by $(\boldsymbol{\delta}^*)' = \boldsymbol{\delta}'P$. Then

$$\Delta_1 \leqslant (1 - \lambda)\, \Delta_0$$

where
$$0 \leqslant \lambda = \max_j \{\min_i p_{ij}\}.$$

Proof. First note that $(\boldsymbol{\delta}^*)'\mathbf{1} = \boldsymbol{\delta}'P\mathbf{1} = \boldsymbol{\delta}'\mathbf{1} = 0$. If $\boldsymbol{\delta}^* = \mathbf{0}$, the assertion is trivial. Hence assume $\boldsymbol{\delta}^* \neq \mathbf{0}$.

Permute the rows and columns of P simultaneously in such a way that λ is the uniform lower bound for the *first* column; this only permutes the elements of $\boldsymbol{\delta}$ and $\boldsymbol{\delta}^*$ in a similar fashion (and does not change λ). Hence we can assume without loss of generality that $p_{i1} \geqslant \lambda$, $i = 1, \ldots, n$. Also, by multiplying both $\boldsymbol{\delta}$ and $\boldsymbol{\delta}^*$ by -1 if necessary, we can also assume without loss of generality that $\delta_1{}^* \leqslant 0$.

Finally, let $\delta_i^+ = \delta_i$ if $\delta_i > 0$, $= 0$ if $\delta_i \leqslant 0$; and similarly for δ_i^*.

Clearly

$$\frac{1}{2}\Delta_1 = \sum_{i=1}^{n} (\delta_i^*)^+, \frac{1}{2}\Delta_0 = \sum_{i=1}^{n} \delta_i^+.$$

Further, since

$$\delta_j^* = \sum_{i=1}^{n} \delta_i\, p_{ij}$$

$$\frac{1}{2}\Delta_1 = \sum_{j=1}^{n} (\delta_j^*)^+ = \sum_{i=1}^{n} \delta_i \sum_j{}' p_{ij}$$

where the sumation Σ' is only over the j for which $\delta_j^* > 0$.

$$\leqslant \sum_{i=1}^{n} \delta_i^+ \sum_j{}' p_{ij}$$

$$\leqslant \sum_{i=1}^{n} \delta_i^+ \sum_{j=2}^{n} p_{ij}, \text{ since } \delta_1{}^* \leqslant 0,$$

$$= \sum_{i=1}^{n} \delta_i^+ (1 - p_{i1})$$

$$\leqslant (1 - \lambda) \sum_{i=1}^{n} \delta_i^+$$

since $p_{i1} \geqslant \lambda$, for all i

$$= (1 - \lambda)\,(1/2)\Delta_0$$

so that the proof is complete. #

This result leads to the following definition, important for the sequel:

DEFINITION 4.6. A Markov matrix is a stochastic matrix with positive λ, i.e. a stochastic matrix with at least one column entirely positive.

The set of Markov matrices is a subset of the class of regular stochastic matrices†, defined in the previous section. We henceforth refer to the λ appropriate to a stochastic matrix P by $\lambda(P)$.

The following theorem, proved with the aid of Lemma 4.1, is the oldest, and still most fundamental (as we shall see from the sequel) result on finite inhomogeneous Markov chains.

THEOREM 4.8. For a sequence $S = \{P_i\}$ of $(n \times n)$ stochastic matrices, there is tendency to stability of the inhomogeneous products $T_{p,k}$ as $k \to \infty$, $p = 0, 1, 2, \ldots$ (i.e. weak ergodicity obtains) if

$$\sum_{i=1}^{\infty} \lambda(P_i)$$

diverges.

Proof. Assume for the moment that $\lambda(P_i) < 1$ for all i. We shall deal with the case $p = 0$ first:

$$t_{i,s}^{(0,k+1)} - t_{j,s}^{(0,k+1)} = \sum_{r=1}^{n} (t_{i,r}^{(0,k)} - t_{j,r}^{(0,k)})\, p_{rs}\,(k+1).$$

Consider i and j fixed, $i \neq j$, the above equation holds for $s = 1, \ldots, n$, so that if we put

$$\delta(k) = \{t_{i,s}^{(0,k)} - t_{j,s}^{(0,k)}\},\ k = 0, 1, 2, \ldots$$

we have
$$\delta'(k+1) = \delta'(k)\, P_{k+1}$$

† See Exercise 4.12, to this section.

where $\delta'(k)\,1 = 0$, through stochasticity of $T_{0,k}$ for all k. Putting

$$\Delta(k) = \sum_{i=1}^{n} |\delta_i(k)|,$$

we have by Lemma 4.1

$$\Delta(k+1) \leqslant (1 - \lambda(P_{k+1}))\,\Delta(k),\ k = 0, 1, 2, \ldots$$

Therefore
$$\Delta(k) \leqslant 2 \prod_{i=1}^{k} (1 - \lambda(P_i)),\ k \geqslant 1,$$

since $\Delta(0) = 2$. Hence $\Delta(k) \to 0$ as $k \to \infty$ if

$$\prod_{i=1}^{\infty} (1 - \lambda(P_i)) = 0,$$

i.e. if $\sum_{i=1}^{\infty} \lambda(P_i) = \infty$.

The same argument goes through in essence for any p, since the tails of the above infinite product and sum are involved, and these are the operative quantities.

We leave the case where $\lambda(P_i) = 1$ for some i to an exercise.† #

COROLLARY 1. For all $i, j, s = 1, \ldots, n$: $p \geqslant 0$ and $k \geqslant 1$:

$$|t_{i,s}^{(p,k)} - t_{j,s}^{(p,k)}| \leqslant 2 \prod_{i=p+1}^{p+k} (1 - \lambda(P_i)).$$

COROLLARY 2. If all matrices P_i are uniformly Markov (i.e. $\lambda(P_i) \geqslant \lambda_0 > 0$ for all i), then weak ergodicity obtains at a rate which is at least geometric with rate $(1 - \lambda_0)$.

Theorem 4.8 does not restrict, in any obvious way, the structure of all component stochastic matrices, $S = \{P_i\}$, entering into the inhomogeneous products $T_{p,k}$ which we wish to consider as $k \to \infty$. In the remaining discussion on weak ergodicity, *we shall assume that each P_i is regular*; and for convenience call the class of all $n \times n$ regular matrices G_1. We denote the class of all $n \times n$ Markov matrices (defined above) by M, and note $M \subset G_1$. Finally, we define at this stage for convenience, an 'intermediate' class G_2, such that $M \subset G_2 \subset G_1$, by allocating an $n \times n$ stochastic matrix P to G_2 if (i) $P \in G_1$, (ii) $QP \in G_1$, for any $Q \in G_1$.

We shall write $P_A \sim P_B$ for any two $(n \times n)$ stochastic matrices P_A and P_B if they have *common incidence matrix* i.e. if they have zero elements and positive elements in the same positions, so that the 'pattern' is the same. Whether or not a matrix $P \in G_1$, G_2 or M clearly depends only on its

† See Exercise 4.13. By consideration of this situation the reason for defining weak ergodicity in terms of the behaviour of $T_{p,k}$ as $k \to \infty$ for *each* p, rather than of $T_{0,k}$ alone, becomes clear.

incidence matrix. For $n \times n$ matrices, there is clearly only a finite number of possible incidence matrices (= Boolean relation matrices = 'patterns').

LEMMA 4.2. If P and Q are stochastic, and Q or $P \in M$, htne PQ and $QP \in M$.

Proof. See Exercise 4.14. #

LEMMA 4.3. If $Q \in G_1$, P is stochastic, and $PQ \sim P$, then $P \in M$.

Proof. If $PQ \sim P$, then $PQ^2 = (PQ)Q \sim PQ \sim P$; and similarly for every positive integer k, $PQ^k \sim P$. But for k sufficiently large $Q^k \in M$, since $Q \in G_1$, by Theorem 4.6. Hence, by Lemma 4.2, PQ^k, and hence P, $\in M$. #

LEMMA 4.4. If P and $Q \in G_2$, then so do PQ and QP.

Proof. Let R be any matrix of G_1. Then

$$R(PQ) = (RP)Q$$

where $RP \in G_1$ since $P \in G_2$. Hence since $Q \in G_2$, $R(PQ) = (RP)\,Q \in G_1$. Further since $P \in G_2 \subset G_1$, and $Q \in G_2$, $PQ \in G_1$. Thus $PQ \in G_2$.

A similar argument holds for QP. #

This lemma shows G_2 is closed under multiplication, though it is easily seen G_1 is not.†

LEMMA 4.5. If for each $p \geqslant 0$, $T_{p,k} \in G_1$ for all $k \geqslant 1$, then $T_{p,k} \in M$ for $k \geqslant t+1$, where t is the number of distinct types of incidence matrices corresponding to G_1.

Proof. Consider $T_{p,t+1}$ for any fixed $p \geqslant 0$, and put $B_i = T_{p+i}$ for convenience, so that $T_{p,t+1} = \prod_{i=1}^{t+1} B_i$.

Now, from the definition of t, there exist integers a and b, $0 < a < b \leqslant t+1$ such that

$$B_1B_2 \ldots B_c \sim B_1B_2 \ldots B_b$$

since among the $t+1$ quantities $\prod_{i=1}^{j} B_i, j = 1, 2, \ldots, t+1$ taken in sequence, one of the t possible patterns (incidence matrices) must eventually recur. Hence, since $B_{a+1} \ldots B_b \in G_1$ by assumption, Lemma 4.3 implies $B_1 \ldots B_a \in M$, so that

$$T_{p,t+1} = B_1B_2 \ldots B_{t+1} \in M$$

by Lemma 4.2 and hence $T_{p,k} \in M$ for $k \geqslant t+1$ by the same result. #

COROLLARY. The assumption that $T_{p,k} \in G_1$ for all k and p may be replaced by the (stronger) assumption that $P_i \in G_2$ for all i, made on the *individual* matrices. (This follows from Lemma 4.4.)

† See Exercise 4.15.

THEOREM 4.9. If for each $p \geqslant 0$, $T_{pk} \in G_1$ for all $k \geqslant 1$ and

$$\min_{i,j}{}^{+} p_{ij}(k) \geqslant \delta > 0$$

uniformly for all $k \geqslant 1$, weak ergodicity obtains (uniformly for all p).

Proof. First note that

$$T_{0,k(t+1)} = T_{0,t+1}\, T_{t+1,t+1}\, T_{2(t+1),t+1} \cdots T_{(k-1)(t+1),t+1}$$

where $T_{i(t+1),t+1}$, $i \geqslant 0$, is Markov, in virtue of Lemma 4.5. Moreover the condition on the individual component P_is ensures that $\lambda(T_{i(t+1),\, t+1}) \geqslant \delta^{t+1}$, so that, by Corollary 2 of Theorem 4.8, weak ergodicity obtains at least for the sequence $\{T_{i(t+1),t+1}\}$, $i \geqslant 0$.

A totally analogous argument obtains for any sequence $\{T_{p+i(t+1),t+1}\}$, $i \geqslant 0$, for any fixed $p \geqslant 0$.

Now, consider any fixed p, and k 'large'; then

$$T_{p,k} = T_{p,t+1}\, T_{p+(t+1),t+1} \cdots T_{p+r(t+1),t+1}\, \bar{T}_{(k)}$$

where $r \equiv r(k)$ is the largest positive integer such that $(r+1)(t+1) \leqslant k$ and $\bar{T}_{(k)}$ is some stochastic matrix. Using Corollary 1 of Theorem 4.8 for all $i, j, s = 1, \ldots, n$

$$|t_{i,s}^{(p,k)} - t_{j,s}^{(p,k)}| \leqslant 2 \prod_{i=0}^{r(k)} (1 - \lambda(T_{p+i(t+1),(t+1)}))$$

since $1 - \lambda(\bar{T}_{(k)}) \leqslant 1$;

$$\leqslant 2(1 - \delta^{t+1})^{r(k)} \to 0$$

as $k \to \infty$, since $r(k) \to \infty$; and this complete the proof. #

We conclude the discussion of weak ergodicity with a discussion of the nature of matrices of the class G_2; naturally, one prefers a condition on the individual P_is rather than on the $T_{p,k}$s in applying Theorem 4.9: and Lemma 4.4 for example ensures that if all $P_i \in G_2$, then this will suffice in the context of the theorem. The G_2 condition is not easy to check, and to the ultimate end of practical application, we introduce another definition.

DEFINITION 4.7. An $(n \times n)$ stochastic matrix $P = \{p_{ij}\}$ will be called a *scrambling* matrix, if given two rows α and β, there is at least one column, γ, such that $p_{\alpha\gamma} > 0$ and $p_{\beta\gamma} > 0$.

Clearly Markov matrices are scrambling; but regular matrices are not necessarily so.†

LEMMA 4.6. A scrambling matrix $P \in G_2$.

Proof. Consider the matrix P in canonical form, so that its essential classes

† See Exercise 4.18.

of indices are also in canonical form if periodic. (Clearly the scrambling property is invariant under simultaneous permutation of rows and columns.)

It is now easily seen that if there is more than one essential class the scrambling property fails by judicious selection of rows α and β in different essential classes; and if an essential class is periodic, the scrambling property fails by choice of α and β in different cyclic subclasses. Thus $P \in G_1$.

Now consider a scrambling $P = \{p_{ij}\}$, not necessarily in canonical form. Then for any stochastic $Q = \{q_{ij}\}$, QP is scrambling, for take any two rows α,β and consider the corresponding entries

$$\sum_{k=1}^{n} q_{\alpha k}\, p_{kj},\ \sum_{r=1}^{n} q_{\beta r}\, p_{rj}$$

in the jth column of QP. Then there exist k, r such that $q_{\alpha k} > 0$, $q_{\beta r} > 0$ by stochasticity of Q. By the scrambling property of P, there exists j such that $p_{kj} > 0$ and $p_{rj} > 0$. Hence QP is scrambling, and so $QP \in G_1$ by the first part of the theorem.

Hence, putting both parts together, $P \in G_2$. #

COROLLARY. For *any* stochastic Q, QP is scrambling, for fixed scrambling P.

This corollary motivates the following result, which generalizes Lemma 4.2.

LEMMA 4.7. If P is scrambling matrix, then so are PQ and QP for any stochastic Q.

Proof. In view of the preceding, it is necessary to prove only that PQ is scrambling for every stochastic Q. Since $P = \{p_{ij}\}$ is scrambling, for any pair of indices (i_1, i_2) there is a $j = j(i_1, i_2)$ such that $p_{i_1,j} > 0, p_{i_2,j} > 0$. Now there is a k such that $q_{jk} > 0$, by stochasticity of Q. Hence the kth column of PQ has positive entries in the i_1th and i_2th rows, as required. #

The conclusion of this result can be strengthened.

LEMMA 4.8. For fixed P, $PQ \in G_1$ for every stochastic Q if and only if P is scrambling.

Proof. If P is scrambling, PQ is scrambling by Lemma 4.7, so $PQ \in G_2 \subset G_1$ (by Lemma 4.6).

Suppose P is not scrambling; then there exist rows α, β such that for each j, $p_{\alpha j} > 0$ implies $p_{\beta j} = 0$ and similarly with α, β interchanged, although by stochasticity of P, each row contains at least one positive entry.

Let Q be a specific stochastic matrix with all columns zero, except the αth and βth which are defined as follows

$$\begin{aligned} q_{i\alpha} &= 1 \text{ if } p_{\beta i} > 0 \\ &= 0 \text{ if } p_{\beta i} = 0; \\ q_{i\beta} &= 1 \text{ if } p_{\alpha i} > 0 \\ &= 0 \quad \text{otherwise.} \end{aligned}$$

It follows that $R = PQ$ has all columns zero except αth and βth, and that the αth and βth rows of $R = \{r_{ij}\}$ have as their only positive entries

$$r_{\alpha\beta} = 1 = r_{\beta\alpha}.$$

It thus follows that in R, the pair of indices $\{\alpha, \beta\}$ form an essential class of period 2. Hence $R \notin G_1$. #

Let us at this stage denote by G_3 the class of $(n \times n)$ scrambling matrices. It follows from the above that

$$M \subset G_3 \subset G_2 \subset G_1$$

and that matrices of G_3 have two special properties: (i) it is easy to verify whether or not a matrix $\in G_3$; (ii) if a single matrix $P_i \in G_3$ is present in $T_{p,k}$ for some fixed p, and all k sufficiently large, then $T_{p,k} \in G_1$ for all k sufficiently large (from Lemma 4.7). This is especially useful if one adopts (as is sometimes done) the (weaker) version of the weak ergodicity definition pertaining only to $T_{0,k}$; for then a single $P_i \in G_3$ ensures that $T_{0,k} \in G_1$ for all $k \geqslant i$.

However, as regards the present definition of weak ergodicity, an *infinite subsequence of scrambling matrices* of the sequence $\{P_i\}$ is required towards the same end.

Finally we mention the important result that if the sequence $\{P_i\}$ has each of its elements selected from a numerically finite set of stochastic matrices, and is such that $T_{p,k} \in G_1$ for all p, k, then weak ergodicity obtains (uniformly). This is a direct consequence of Theorem 4.9, since the 'boundedness from zero' condition in the present context is automatically satisfied.

There remains the slightly vexing question: is G_3 a *proper* subject of G_2; at least for matrices $n \times n$ with n large enough? In other words, are there n, such that $P \in G_2$ but $P \notin G_3$?

The answer is *yes*; we shall give an example when $n = 4$, and for further information refer the reader to Exercises 4.19–4.21.

The stochastic matrix P whose incidence matrix is

$$\tilde{P} = \begin{bmatrix} 1 & 1 & 0 & 0 \\ 1 & 1 & 0 & 0 \\ 1 & 0 & 1 & 0 \\ 0 & 1 & 0 & 1 \end{bmatrix}$$

is a member of G_1 (the first and second indices form an essential aperiodic class; the others are inessential), but not of G_3 (on account of its third and fourth rows). Further, $\tilde{P}$ can be expressed in the form

$$\tilde{P} = I + C$$

where C is an incidence matrix, since $\widetilde{P}$ has positive diagonal elements (such P are said to be 'normed'.) The justification that $P \in G_2$ is completed by the following.

LEMMA 4.9. If P has incidence matrix of form $\widetilde{P} = I + C$, where C is an incidence matrix, and $Q \in G_1$ then QP and $PQ \in G_1$. In particular, $P \in G_2$, if also $P \in G_1$.

Proof.

$$QP \sim Q(I + C) \sim Q + C_2$$

where C_2 is the incidence matrix of QC. Hence indices which were aperiodic in Q, remain so in QP. Moreover, if E_1 denotes the set of essential indices in Q, and E_2 the corresponding set in QP, $E_2 \supset E_1$, and E_2 forms a single essential class, which consequently must be aperiodic.

The case PQ is similar. #

That the set E_1 may actually be a proper subset of E_2 is demonstrated by taking P to be a stochastic matrix with every entry positive.

Thus 'normed' matrices of G_1 are members of G_2, but not necessarily of G_3.

Strong ergodicity

Recall that strong ergodicity requires in addition to weak ergodicity, the elementwise existence of the limit of each $T_{p,k}$, $p \geqslant 0$, as $k \to \infty$, so that $T_{p,k} \to \mathbf{1}\, D_p'$, $p \geqslant 0$, as $k \to \infty$, where D_p is some probability vector.

DEFINITION 4.8. We shall say that the sequence $\{P_i\}$ (or the corresponding inhomogeneous MC) is *asymptotically homogeneous* if there is a probability vector† D such that

$$\lim_{i \to \infty} D'P_i = D'.$$

DEFINITION 4.9. We shall say that the sequence (or chain) is *asymptotically stationary* if there is a probability vector D such that

$$\lim_{k \to \infty} D'T_{p,k} = D', \; p \geqslant 0.$$

We first explore the relations between these concepts.

LEMMA 4.10. An asymptotically stationary sequence is asymptotically homogeneous.

Proof. For each $p \geqslant 0$, the sequence of row vectors $D'T_{p,k}$ approaches D', for some probability vector D. Now

$$D'T_{p,k} = D'T_{p,k-1}\, P_{p+k}.$$

† i.e. $D \geqslant 0$, $D'\mathbf{1} = 1$

Hence letting $D'T_{p,k+1} = D' + E'_{p,k+1}$, where $E'_{p,k+1} \to 0$ as $k \to \infty$,

we have
$$D'T_{p,k} = D'P_{p+k} + E'_{p,k+1}\,P_{p+k}$$
$$\to D' \text{ as } k \to \infty, \text{ for each } p$$

by asymptotic stationarity, since P_{p+k}, being stochastic, is elementwise uniformly bounded as regards p and k. #

LEMMA 4.11. If the sequence $\{P_i\}$ is asymptotically homogeneous
$$\lim_{p\to\infty} D'T_{p,k} = D', \text{ for each } k \geqslant 1.$$

Proof. The proposition is true for $k = 1$, from assumption, since $T_{p,1} = P_{p+1}$. Assume it is true for some $k \geqslant 1$. Then
$$D'T_{p,k+1} = D'T_{p,k}\,P_{p+k+1}$$

and, writing $D'T_{p,k} = D' + E'_{p,k}$, where $E'_{p,k} \to \mathbf{0}$ as $p \to \infty$

(by the induction hypothesis), we have
$$D'T_{p,k+1} = D'P_{p+k+1} + E'_{p,k}\,P_{p+k+1}$$

so that, again using the asymptotic homogeneity, and stochasticity of P_{p+k+1}
$$\lim_{p\to\infty} D'T_{p,k+1} = D'.$$

Thus the proof by induction is complete. #

The following result is an 'inhomogeneous' generalization of Lemma 4.1, which enables us to eventually strengthen our weak ergodicity results to strong ergodicity, under suitable additional assumptions.

LEMMA 4.12. For a given sequence $\{P_i\}$ of stochastic matrices, let
$$\delta'(k+1) = \delta'(k)\,P_{k+1} + r'(k),\ k = 0, 1, 2, \ldots$$

where $\delta'(k)\,\mathbf{1} = 0 = r'(k)\mathbf{1}$.

Then, putting $\Delta(k) = \sum_{i=1}^{n} |\delta_i(k)|$, $\Gamma(k) = \sum_{i=1}^{n} |r_i(k)|$,
$$\Delta(k+1) \leqslant \Delta(0)\prod_{i=1}^{k+1}(1-\lambda(P_i)) + \sum_{i=0}^{k-1}\Gamma(i)\prod_{j=i+2}^{k+1}(1-\lambda(P_j)) + \Gamma(k)$$

for $k \geqslant 0$. $\left(\sum_{i=0}^{-1} \text{ to be interpreted as } 0.\right)$

Proof. By induction. For $k = 0$
$$\delta'(1) - r'(0) = \delta'(0)P_1,$$

where by Lemma 4.1, together with the elementary inequality $|a - b| \geqslant |a| - |b|$
$$\Delta(1) \leqslant \Delta(0)\,(1 - \lambda(P_1)) + \Gamma(0)$$

so the assertion is true for $k = 0$. For arbitrary $k \geqslant 0$

$$\delta'(k+2) - r'(k+1) = \delta'(k+1)P_{k+2}$$

so that similarly

$$\Delta(k+2) \leqslant \Delta(k+1)\left\{1 - \lambda(P_{k+2})\right\} + \Gamma(k+1).$$

Applying the induction hypothesis to $\Delta(k+1)$, the result follows. #

COROLLARY. If all P_i are uniformly Markov, i.e.

$$\lambda(P_i) \geqslant \lambda_0 > 0 \text{ all } i,$$

and also $\boldsymbol{r}(k) \to \mathbf{0}$ elementwise as $k \to \infty$,

$$\Delta(k) \to 0.$$

[In the first place

$$\sum_{i=0}^{k-1} \Gamma(i) \prod_{j=i+2}^{k+1} (1 - \lambda(P_j)) \leqslant \sum_{i=0}^{k-1} \Gamma(i)\,(1-\lambda_0)^{k+i}$$

$$= \sum_{i=0}^{r} \Gamma(i)\,(1-\lambda_0)^{k-1} + \sum_{i=r+1}^{k-1} \Gamma(i)\,(1-\lambda_0)^{k-i}$$

$$\leqslant (1-\lambda_0)^{k-r} \sum_{i=0}^{r} \Gamma(i) + \sum_{i=r+1}^{k-1} \Gamma(i)\,(1-\lambda_0)^{k-i}.$$

Now select $r \equiv r(\epsilon)$ such that $\Gamma(i) < \epsilon$ for $i > r$; and make use of the fact that $\Gamma(i) < C \equiv$ const. for all i. It follows that by taking k sufficiently large, the right hand side will be arbitrarily small by judicious choice of ϵ.]

LEMMA 4.13. For a sequence $\{P_i\}$ of stochastic matrices which are uniformly Markov, asymptotic stationarity and asymptotic homogeneity are equivalent.

Proof. On account of Lemma 4.10, we need to prove that for the present circumstance asymptotic homogeneity implies asymptotic stationarity.

By the assumption of asymptotic homogeneity, there is a probability vector $\boldsymbol{D}$ such that

$$\boldsymbol{D}'P_i = \boldsymbol{D}' - \boldsymbol{e}'(i)$$

where $\boldsymbol{e}(i) \to \mathbf{0}$ as $i \to \infty$, and $\boldsymbol{e}'(i)\,\mathbf{1} = 0$. Let us write for $k \geqslant 1$

$$\boldsymbol{D}'P_k = \boldsymbol{D}'T_{0,k} + \delta'(k).$$

Then
$$\boldsymbol{D}'P_{k+1} = \boldsymbol{D}'T_{0,k+1} + \delta'(k+1)$$

but
$$\boldsymbol{D}'P_k = \boldsymbol{D}' - \boldsymbol{e}'(k)$$

so that
$$\boldsymbol{D}'P_{k+1} = (\boldsymbol{D}'T_{0,k} + \delta'(k) + \boldsymbol{e}'(k))\,P_{k+1}$$
$$= \boldsymbol{D}'T_{0,k+1} + \delta'(k)\,P_{k+1} + \boldsymbol{e}'(k)\,P_{k+1}.$$

Thus we have for $k \geqslant 0$

$$\delta'(k+1) = \delta'(k)\, P_{k+1} + r'(k)$$

where
$$r'(k) = e'(k)\, P_{k+1},$$

and so
$$r'(k)\mathbf{1} = 0 = \delta'(k)\mathbf{1}$$

(all by stochasticity).

Hence by the Corollary to Lemma 4.12, since

$$r'(k) = e'(k)\, P_{k+1} \to \mathbf{0}' \text{ as } k \to \infty, \text{ since } e'(k) \to \mathbf{0}',$$

if follows that
$$\lim_{k\to\infty} D'P_k = \lim_{k\to\infty} D'T_{0,k} = D'.$$

The same proposition follows for $D'T_{p,k}$ as $k \to \infty$, similarly, since the assumptions on the sequence $\{P_i\}$ are invariant under shift:

$$\lim_{k\to\infty} D'P_{k+p} = \lim_{k\to\infty} D'P_k = \lim_{k\to\infty} D'T_{p,k} = D'. \quad \#$$

THEOREM 4.10. Under the conditions of Theorem 4.9 on the sequence $\{P_i\}$, and the additional condition that the sequence be asymptotically homogeneous, strong ergodicity obtains.

Proof. Since from Theorem 4.9 we know weak ergodicity obtains, we have only to prove the elementwise existence of the limit of each $T_{p,k}$, $p \geqslant 0$, as $k \to \infty$.

First of all we note that if $\mathbf{\Pi} = \{\Pi_i\}$ and $D = \{d_i\}$ are *any two* fixed probability vectors, then the arguments of both Theorems 4.8 and 4.9 and Corollaries go through in terms of

$$\left|\sum_{i=1}^{n} \Pi_i\, t_{i,s}^{(p,k)} - \sum_{i=1}^{n} d_i\, t_{i,s}^{(p,k)}\right|, \; s = 1, \ldots, n$$

rather than the more restricted (fixed initial indices: i, j) situation:

$$|t_{i,s}^{(p,k)} - t_{j,s}^{(p,k)}|, \; s = 1, \ldots, n,$$

except that $\Delta(0) = 2$ of the former situation would need to be replaced by

$$\Delta(0) = \sum_{r=1}^{n} |\Pi_r - d_r|,$$

so that in any case $\Delta(0) \leqslant 2$ (so in fact everything goes through).

In view of this, it is clearly adequate to show that for some probability vector $D = \{d_i\}$,

$$D'T_{p,k} \to D' \text{ as } k \to \infty$$

for all p, since then for all fixed i, p,

$$t_{i,s}^{(p,k)} \to d_s, \; s = 1, 2, \ldots, n$$

as required, so that, elementwise

$$T_{p,k} \to \mathbf{1}\, D'.$$

We take as our vector D the vector guaranteed by asymptotic homogeneity of $\{P_i\}$, such that

$$\lim_{i\to\infty} D'P_i = D'.$$

Consider now the sequence (for p fixed)

$$\{T_{p+i(t+1),t+1}\},\ i \geqslant 0$$

where t is fixed and has the meaning of Lemma 4.5; this is uniformly Markov by the argument of Theorem 4.9; and by Lemma 4.11

$$\lim_{i\to\infty} D'\, T_{p+i(t+1),t+1} = D'$$

so that the sequence

$$T_{p+i(t+1),t+1},\ i \geqslant 0,$$

is asymptotically homogeneous. Thus by Lemma 4.13 the sequence is asymptotically stationary: in particular

$$\lim_{r\to\infty} D'T_{p,t+1}\, T_{p+(t+1),t+1} \cdots T_{p+r(t+1),t+1} = D'.$$

Now, for fixed m, $1 \leqslant m < (t+1)$

$$\lim_{r\to\infty} D'P_{p+(r+1)(t+1)+1}\, P_{p+(r+1)(t+1)+2} \cdots P_{p+(r+1)(t+1)+m}$$

$$= \lim_{r\to\infty} D'\,T_{p+(r+1)(t+1),m}$$

$$= D'$$

by Lemma 4.11. Hence, without difficulty, for fixed m, $0 \leqslant m < t+1$

$$\lim_{r\to\infty} D'T_{p,t+1}\, T_{p+(t+1),t+1} \cdots T_{p+r(t+1),t+1}\, T_{(m,r)}$$

$$= D'$$

where

$$T_{(m,r)} = P_{p+(r+1)(t+1)+1} \cdots P_{p+(r+1)(t+1)+m}, \quad m > 0$$

$$(= T_{p+(r+1)(t+1),m} \qquad m > 0)$$

$$= I \qquad m = 0.$$

Thus

$$D'\,T_{p,(r+1)(t+1)+m} \to D'$$

as $r \to \infty$ for each fixed m, $0 \leqslant m < t+1$. Thus

$$\lim_{k\to\infty} D'\,T_{p,k} = D', \text{ as required.} \quad \#$$

LEMMA 4.14. Suppose strong ergodicity obtains for the sequence $\{P_i\}$ so that

$$\lim_{k\to\infty} T_{p,k} = \mathbf{1}D'_p\,,\ p \geqslant 0,$$

where D_p is a probability vector. Then $D_0 = D_1 = D_2 = \ldots = D_p = D$ say.

Proof.

$$T_{p,k+1} = T_{p,k}\,P_{p+k+1}$$
$$= (\mathbf{1}\,D'_p + E_{p,k})\,(P_{p+k+1})$$

where $E_{p,k} \to 0$ elementwise as $k \to \infty$. Hence letting $k \to \infty$, since P_{p+k+1} is stochastic

$$\mathbf{1}\,D'_p = \lim_{k\to\infty} \mathbf{1}\,D'_p\,P_{p+k+1}$$

so that

$$D'_p = \lim_{k\to\infty} D'_p\,P_k$$

so that the sequence $\{P_i\}$ is asymptotically homogeneous with any one of D_r, $r \geqslant 0$. Hence by Lemma 4.11

$$\lim_{p\to\infty} D'_r\,T_{p,k} = D'_r, \qquad r = 0, 1, 2, \ldots$$

But $T_{p,k} \to \mathbf{1}\,D'_p$ as $k \to \infty$, so that

$$D'_r\,\mathbf{1}\,D'_p = D'_r$$

i.e. $D'_p = D'_r$ for arbitrary r, $p \geqslant 0$,

since $D'_r\,\mathbf{1} = 1$, as D_r is a probability vector. #

THEOREM 4.11. Suppose weak ergodicity obtains for the sequence $\{P_i\}$. Then a necessary and sufficient condition for strong ergodicity is asymptotic stationarity of $\{P_i\}$.

Proof. Suppose strong ergodicity obtains, and D is the probability vector of Lemma 4.14. Then for any $p \geqslant 0$

$$\lim_{k\to\infty} D'T_{p,k} = D'\,\mathbf{1}\,D' = D'$$

so that asymptotic stationarity obtains.

Suppose asymptotic stationarity obtains with the probability vector $D = \{d_i\}$. Then fixed i and p, $s = 1, \ldots, n$:

$$|t_{i,s}^{(p,k)} - \sum_{j=1}^{n} d_j\,t_{j,s}^{(p,k)}| = |\sum_{j=1}^{n} d_j\,\{t_{i,s}^{(p,k)} - t_{j,s}^{(p,k)}\}|$$
$$\leqslant \sum_{j=1}^{n} |t_{i,s}^{(p,k)} - t_{j,s}^{(p,k)}| \to 0$$

as $k \to \infty$ by weak ergodicity. On the other hand

$$\sum_{j=1}^{n} d_j\,t_{j,s}^{(p,k)} \to d_s$$

by asymptotic stationarity. Hence using the elementary inequality $|a - b| \geqslant |a| - |b|$, we obtain

$$\lim_{k\to\infty} |t_{i,s}^{(p,k)} - d_s| \to 0$$

i.e. $$T_{p,k} \to 1\,D' \quad \text{elementwise, as required.} \quad \#$$

THEOREM 4.12. If in the sequence $\{P_k\}$, all $P_k = P = \{p_{ij}\}$ independent of k, weak and strong ergodicity are equivalent.

Proof. We need to prove that weak ergodicity implies strong ergodicity. [We note that

$$t_{i,s}^{(p,k)} = p_{i,s}^{(k)},\ t_{j,s}^{(p,k)} = p_{j,s}^{(k)}$$

so by weak ergodicity

$$|p_{i,s}^{(k)} - p_{j,s}^{(k)}| \to 0 \text{ as } k \to \infty$$

for all fixed i, j, s.]

Now P contains at least one essential class of indices: let D be the probability vector which contains zeroes except in the positions corresponding to the indices of the essential class, where the entries are those of the unique stationary distribution corresponding to this class:† then $D'P = D'$, and indeed $D'P^k = D'$, so that the sequence $\{P_i\}$ is asymptotically (in fact exactly) both homogeneous and stationary.

Hence by Theorem 4.11 the result follows. #

COROLLARY. In the circumstances of the theorem, asymptotic homogeneity and asymptotic stationarity are equivalent.

Bibliography and discussion to §4.3

The definition of weak ergodicity is due to Kolmogorov (1931), who proves weak ergodicity for a sequence $\{P_i\}$ of finite or infinite stochastic matrices, under a rather restrictive condition not discussed here. (See also Sarymsakov (1954), §4, for a repetition of the reasoning.) Lemma 4.1 and the reasoning of Theorem 4.8 and its Corollaries goes back to Markov (1907) (see also Markov (1924)) although in the present degree of generality the theorem is due to Bernstein (1946) (see also Bernstein (1964) for a reprinting of the material). The theorem was rediscovered by Mott (1957). It is known in the Russian literature as Markov's Theorem.

As regards the remainder of the weak ergodicity theory given, the introduction of the class G_2, Lemmas 4.2 to 4.5, and Theorem 4.9 are all due to Sarymsakov (1953*a*; summary) and Sarymsakov & Mustafin (1957), although the simple proof of Lemma 4.5 as presented here is essentially that of Wolfowitz (1963). Weaker versions of Theorem 4.9, pertaining to situation where each matrix P_i is chosen from a finite set, were obtained by Wolfowitz

†See Theorem 4.1.

(1963) and Paz (1963, 1965). The substantial insight into the problem due to Sarymsakov and Mustafin consists in noting that the theory can be developed in terms of the (at first apparently restrictive) notion of a Markov matrix, and made to depend, in the end, on Markov's Theorem, a remarkable situation if one takes into account the age of Markov's Theorem.

In the west, study of weak (and strong) ergodicity was initiated by Hajnal (1956, 1958), to whom, and to Sarymsakov (1956, 1958) the elegant notion of scrambling matrix, and thus of the class $G_3 \subset G_2$, is due. Lemmas 4.7 and 4.8 are proved by Hajnal within the later of his cited papers†; the work of Mott, Wolfowitz, and Paz (1965) already cited follows from that of Hajnal, which is developed in terms of a coefficient, or measure, of ergodicity; this notion has been avoided in the present development (although our 'column measure' $\lambda(P)$ of a stochastic P in some degree fulfils the same purpose). Lemma 4.9 is again due to Sarymsakov & Mustafin (1957).

As regards the strong ergodicity theory presented, the notion of asymptotic homogeneity is again due to Bernstein (1946, 1964) as are, in essence, Lemmas 4.12 and 4.13. The concept of asymptotic stationarity appears to be due to Kozniewska (1962), as is Lemma 4.10. Lemma 4.11 and Theorem 4.10 (the strong ergodicity extension of the Sarymsakov–Mustafin Theorem) in the present degree of generality, are given in Seneta (1973a), although for uniform Markov matrices $\{P_i\}$ the extension was given by Bernstein (1946) (see also Mott (1957)). Theorems 4.11 and 4.12 are due to Kozniewska (1962).

For results not included in the present treatment, but related to it, including necessary and sufficient conditions for weak ergodicity, we refer the reader to the papers of Sarymsakov, Mustafin, Hajnal, Mott and Kozniewska already cited, and also to Mott & Schneider (1957). Possibly the first instance of a necessary and sufficient condition for weak ergodicity occurs in the paper of Doeblin (1937), who obtains it from an earlier sufficient condition of Ostenc (1934) – see Seneta (1973b).

We conclude the present discussion with some notes on results on non-homogeneous finite Markov chains a little outside the scope of the present treatment.

Although all our theoretical development has been for a sequence $S = \{P_i\}$ of stochastic matrices which are of the same row and column dimensions, viz. $(n \times n)$, we have mentioned earlier (in connection with the Pólya Urn scheme) that some inhomogeneous Markov chains do not have a constant state space; and in general one would look at the case where in the sequence S, if P_i is $n_i \times n_{i+1}$, the P_{i+1} is $n_{i+1} \times n_{i+2}$, $i \geqslant 1$. In fact it is possible to carry some of the theory through, in the manner of Bernstein (1946), for this situation (in particular Lemma 4.1 and Theorem 4.8 are easily extended to this situation). Some more recent writing on finite inhomogeneous Markov chains of the Russian school also pertains to this more general framework e.g. Sarymsakov (1958, 1961).

† Lemmas 4.6 and 4.7 were announced by Sarymsakov (1956) without proof.

The more restricted situation, also mentioned several times, where each of the $n \times n$ matrices P_i, $i \geqslant 1$, is chosen from some fixed finite set K is of relevance to coding (information) theory. We have seen that a fundamental assumption in our theorems is that $T_{p,k} \in G_1$, $p \geqslant 0, k \geqslant 1$. In the coding context, where one wishes to consider not just *one* sequence $S = \{P_i\}$, with the P_is chosen as stated, but *all possible sequences* which can be so formed, the corresponding requirement is that all 'words' in the member matrices of the finite fixed set be regular. The validity (or not) of this assumption may be verified in a finite number of operations, and substantial effort has been dedicated to obtaining bounds on the amount of labour required. A best possible result in this connection is due to Paz (1965), who shows that one needs only to check all possible words of length†

$$\nu \leqslant \left(\frac{1}{2}\right)(3^n - 2^{n+1} + 1);$$

for some such ν, all words of length ν will be scrambling, if and only if all possible words are regular. Earlier and weaker bounds of this sort were obtained by Thomasian (1963) and Wolfowitz (1963). In fact we may see from the *proof* of Lemma 4.5, with the same reasoning as Wolfowitz, that if all words of length $\nu \leqslant t$ are regular, then all possible words of length $t + 1$ are Markov, so that *all* possible words are regular. It is clear that, in this context, a condition on the individual matrices of the finite fixed set K (from which the P_is are chosen) which would *ensure* that all words are regular would be most desirable: Lemma 4.4 shows that if $K \subset G_2$, then this is so. A substantial, but not wholly conclusive, study of the class G_2 occurs in the paper of Sarymsakov & Mustafin (1957), as do also studies of related classes of matrices. Finally, we mention that Sarymsakov (1958, 1961) has given conditions on the individual members of the set K, such that any word of length $(n - 1)$ in these matrices is scrambling, where all matrices considered are $(n \times n)$. The reader will recall that any word containing a scrambling factor is scrambling, and so regular (Lemmas 4.6, 4.7). Also, we mention in this connection that there is a large information-theoretic literature pertaining to this problem: some leads to this may be found in the references cited.

It remains to mention that, as regards the question of weak ergodicity beyond the overall framework of the regular class G_1, some investigations have been made by Larisse & Schützenberger (1964): see also Larisse (1966).

To conclude, it is almost unnecessary to point out that there is a very large literature pertaining to probabilistic aspects of non-homogeneous finite Markov chains other than weak and strong ergodicity, e.g. the Central Limit Theorem and Law of the Iterated Logarithm. This work has been carried on largely by the Russian school; a comprehensive recent reference list to it may be found in Sarymsakov (1961), and an earlier one in Doeblin (1937). Much of the early work is again due to Bernstein (see Bernstein (1964) and other papers in the same book).

† See Exercise 4.23 for the definitions of these terms.

Exercises on §4.3

4.12. Show that a Markov matrix has only one essential class of indices, which is aperiodic, though there may be some inessential indices. Thus a Markov matrix is *regular.*

4.13. Prove Theorem 4.8 when $\lambda(P_i) = 1$ for at least one $i \geqslant 1$.

4.14. Prove Lemma 4.2.

4.15. Show by example that the set G_1 is not closed under multiplication. Show, however, that if $P \in G_1$, $Q \in G_1$ then either both or neither of PQ, $QP \in G_1$.

4.16. Show by example that it is possible that $A, B, C \in G_1$, such that AB, BC, $AC \in G_1$, but $ABC \notin G_1$, although $BAC \in G_1$. (Contrast with the result of Exercise 4.15.)

(Sarymsakov, 1953*a*)

4.17. Suppose that weak ergodicity obtains for a sequence $\{P_i\}$ of stochastic matrices, not necessarily members of G_1. Show that for each fixed $p \geqslant 0$, there exists a strictly increasing sequence of integers $\{m_i\}$, $i \geqslant 1$, such that

$$T_{m_i, m_{i+1}} \in M, \; i \geqslant 0$$

where $m_0 = p$.

(Sarymsakov (1953*a*); Sarymsakov & Mustafin (1957))

Hint: A row of an $n \times n$ stochastic matrix has at least one entry $\geqslant n^{-1}$.

4.18. Show by example that a regular stochastic matrix is not necessarily scrambling.

4.19. Discuss the relation between M, G_2 and G_3 when the dimensions of the stochastic matrices are $n \times n$, where $n = 2, 3$. Discuss the relation between G_2 and G_3 when $n \geqslant 5$.

4.20. Show by examples that a Markov matrix is not necessarily a 'normed' matrix of G_1 (i.e. a matrix of G_1 with positive diagonal); and vice versa. Thus neither of these classes contains the other.

4.21. Show that, for $n = 4$, a scrambling matrix $P = \{p_{ij}\}$ is 'nearly Markov', in that there is a column j such that $p_{i_1,j} > 0$, $p_{i_2,j} > 0$, $p_{i_3,j} > 0$ for distinct i_1, i_2, i_3. Extend to $n > 4$.

Hint: For an $n \times n$ scrambling matrix there are $n(n-1)/2$ distinct pairs of (row) indices, but only n actual (column) indices.

4.22. Show that if P^k is scrambling for some positive integer k, then $P \in G_1$.

(Paz, 1963)

4.23. Let $P_1, \ldots, P_k$ be a finite set of stochastic matrices of the same order. A product of r Ps (repetitions permitted) is called a *word* of *length r.*

Show that if there is an r such that all words in the Ps of length at least r are scrambling, then each word in the $P\text{s} \in G_1$.

Hint: Use Exercise 4.22.

(Paz, 1965)

4.24. Let $\{P_i\}$ be a weakly ergodic sequence of stochastic matrices, and let $\det\{P_i\}$ denote, as usual, the determinant of P_i. Show that

$$\sum_{i=1}^{\infty} (1 - |\det\{P_i\}|) = \infty.$$

(Sirazhdinov, 1950)

4.25. Let us call a stochastic matrix $P = \{P_{ij}\}$ *quasi-Markov* if for a proper subset A of index set $\mathscr{S} = \{1, 2, \ldots, n\}$

$$\sum_{j \in A} p_{ij} > 0, \text{ for each } i \in \mathscr{S}.$$

[A Markov matrix is thus one where A consists of a single index].

Show that a scrambling matrix is quasi-Markov, but (by examples) that a quasi-Markov matrix $\notin G_1$ (i.e. is not regular) necessarily.

Hint: Use the approach of Exercise 4.21.

4.26. Show that if in the sequence $\{P_i\}$, $P_i \to P$ as $i \to \infty$, where P is a regular matrix, strong ergodicity obtains for the sequence.

(Mott, 1957)

Hint: Show that for any fixed p, $T_{p,k}$ is Markov for $k \geqslant k_0$ (where k_0 does not depend on p), since P^r is Markov for $r \geqslant r_0$; and that $\{P_i\}$ is asymptotically homogeneous.

PART II

COUNTABLE NON-NEGATIVE MATRICES

5 Countable stochastic matrices

We initiate our brief study of non-negative matrices with countable index set by a study of stochastic matrices for two main reasons. Firstly the theory which can be developed with the extra stochasticity assumption provides a foundation whose analytical ideas may readily be generalized to countable matrices which are not necessarily stochastic; and this will be our approach in the next chapter. Secondly the theory of countable stochastic matrices is of interest in its own right in connection with the theory of Markov chains on a countable state space; and indeed it is from this framework that the analytical development of our ideas comes, although we shall avoid probabilistic notions apart from the occasional aside. We shall not deal with inhomogeneous situations at all in the interests of brevity, since the infinite matrix theory is as yet of lesser utility than that for finite matrices.

We adopt the obvious notation: we deal with a square (non-negative) stochastic matrix $P = \{p_{ij}\}$ with a countable (i.e. finite or denumerably infinite index set $\{1, 2, \ldots\}$. We recall from §4.1 that the powers $P^k = \{p_{ij}^{(k)}\}$, $k \geqslant 1$, $(P^0 = I)$ are well defined by the natural extension of the rule of matrix multiplication, and are themselves stochastic. [We note in passing that for an arbitrary non-negative matrix T with denumerably infinite index set, some entries of its powers T^k, $k \geqslant 2$ may be infinite, so that the powers are not well defined in the same sense.]

5.1 Classification of indices

Previous classification theory

Much of the theory developed in §1.2 for general finite non-negative matrices goes through in the present case.

In particular the notions of one index leading to another, two indices communicating, and the consequent definition of *essential and inessential* indices, and hence classes, remain valid, as also does the notion of *period* of an index, and hence of a self-communicating class containing the index; and so the notion of *irreducibility* of a matrix P and its index set (see §1.3). A *primitive* matrix P may then be *defined* as a matrix corresponding *to an irreducible aperiodic index set.* Moreover, as before, all these notions depend only on the location of the positive elements in P and not on their size.

In actual fact the only notions of §1.2 which do not necessarily hold in the present context are those concerned primarily with pictorial representation, such as the *path diagram,* and the *canonical form,* since we have to deal now with an index set possibly infinite. Nevertheless, things tend to 'work' in the same way: and consequently it is worthwhile to keep in mind even these notions as an aid to the classification and power-behaviour theory, even though e.g. it may be impossible as regards the canonical form representation to write two infinite sets of indices (corresponding to two self-communicating classes) as one following the other in sequential order.

Of the *results* proved in §§1.2–1.3, viz. Lemma 1.1, Lemma 1.2, Lemma 1.3, Theorem 1.3, Lemma 1.4 and Theorem 1.4, only Lemma 1.1 does not extend to the present context (even Theorem 1.4 remains partly, though no longer fundamentally, relevant). To see that an infinite stochastic matrix with at least one positive entry in each row does not necessarily possess at least one essential class of indices, it is only necessary to consider the example

$$P = \begin{bmatrix} 0 & 1 & 0 & 0 & 0 & \\ 0 & 0 & 1 & 0 & 0 & \cdots \\ 0 & 0 & 0 & 1 & 0 & \\ \vdots & & & & \ddots & \end{bmatrix}$$

i.e.

$$\left.\begin{aligned} p_{ij} &= 1 \text{ for } j = i + 1 \\ &= 0 \text{ otherwise} \end{aligned}\right\} \; i, j \in \{1, 2, \ldots\}.$$

Examples: (1)† Suppose P has the incidence matrix

$$\begin{bmatrix} 1 & 1 & 0 & 0 & 0 & \\ 1 & 0 & 1 & 0 & 0 & \cdots \\ 1 & 0 & 0 & 1 & 0 & \\ \vdots & & & & \ddots & \end{bmatrix}.$$

Then P is irreducible, since $i \to i + 1$ for each i so that $1 \to i$; also $i \to 1$ for each i. Moreover $p_{11} > 0$, so that index 1 is aperiodic. Hence the entire index set $\{1, 2, \ldots\}$ is aperiodic. Consequently P is primitive.

† Relevant to Example (4) of §4.1 if all f_i there are *positive.*

(2) Suppose P has incidence matrix

$$\begin{bmatrix} 1 & 0 & 0 & 0 & 0 & & \\ 1 & 0 & 1 & 0 & 0 & \cdots & \\ 0 & 1 & 0 & 1 & 0 & & \\ 0 & & \ddots & & \ddots & & \\ \vdots & & & & & & \end{bmatrix}.$$

Then (i) index 1 itself forms a single essential aperiodic class. (ii) Each $j \in \{2, 3, \ldots\}$ is inessential, since $j \to j-1$, so that $j \to 1$; however $\{2, 3, \ldots\}$ is a non-essential self communicating class, since also $j \to j+1$. Further the subset has period 2, since clearly for each index $j \in \{2, 3, \ldots\}$ 'passage' is possible only to each of its adjacent indices, so 'return' to any index j can occur at (all) even k in $p_{jj}^{(k)}$, but not at any odd k.

(3) In the counterexample to Lemma 1.1 (in this context) given above, each index is inessential, and each forms a single non-self-communicating inessential class.

New classification theory

Although the above, previously developed, classification theory is of considerable value even in connection with possibly infinite P, it is not adequate to cope with problems of possibly extending Perron–Frobenius theory even to some extent to the present context, and so to deal with the problem of asymptotic behaviour as $k \to \infty$ of $P^k = \{p_{ij}^{(k)}\}$, a fundamental problem insofar as this book is concerned.

It is therefore necessary to introduce a rather more sensitive classification of indices which is used in conjunction with the previous one: specifically, we classify each index as recurrent or transient, the recurrent classification itself being subdivided into positive- and null-recurrent. To introduce these notions, we need additional concepts. Write

$$\ell_{ij}^{(1)} = p_{ij}; \; \ell_{ij}^{(k+1)} = \sum_{\substack{r \\ r \neq i}} \ell_{ir}^{(k)} p_{rj}, \; k \geqslant 1 \tag{5.1}$$

(where $\ell_{ij}^{(0)} = 0$, by definition, for all $i, j \in \{1, 2, \ldots\}$). [In the MC framework, $\ell_{ij}^{(k)}$ is the probability of going from i to j in k steps, without revisiting i in the meantime; it is an example of a 'taboo' probability.]

It is readily seen that $\ell_{ij}^{(k)} \leqslant p_{ij}^{(k)}$ ($\leqslant 1$) by the very definitions of $\ell_{ij}^{(k)}$, $p_{ij}^{(k)}$. Thus for $|z| < 1$, the generating functions

$$L_{ij}(z) = \sum_k \ell_{ij}^{(k)} z^k, \; P_{ij}(z) = \sum_k p_{ij}^{(k)} z^k$$

are well defined for all $i, j = 1, 2, \ldots$

LEMMA 5.1. For $|z| < 1$:

$$P_{ii}(z) - 1 = P_{ii}(z)L_{ii}(z); P_{ij}(z) = P_{ii}(z)L_{ij}(z), i \neq j.$$

Further, $|L_{ii}(z)| < 1$, so that we may write

$$P_{ii}(z) = [1 - L_{ii}(z)]^{-1}.$$

Proof. We shall first prove by induction that for $k \geqslant 1$

$$p_{ij}^{(k)} = \sum_{t=0}^{k} p_{ii}^{(t)}\, \ell_{ij}^{(k-t)}. \tag{5.2†}$$

The proposition is clearly true for $k = 1$, by virtue of the definitions of $\ell_{ij}^{(0)}$, $\ell_{ij}^{(1)}$.

Assume it is true for $k \geqslant 1$.

$$p_{ij}^{(k+1)} = \sum_r p_{ir}^{(k)} p_{rj}$$

$$= \sum_r \Big[\sum_{t=0}^{k} p_{ii}^{(t)}\, \ell_{ir}^{(k-t)}\Big] p_{rj}$$

by the induction hypothesis:

$$= \sum_{t=0}^{k} \Big\{\sum_{r \neq i} \ell_{ir}^{(k-t)}\, p_{rj} + \ell_{ii}^{(k-t)}\, p_{ij}\Big\}\, p_{ii}^{(t)}$$

$$= \sum_{t=0}^{k-1} \big\{\ell_{ij}^{(k-t+1)}\big\}\, p_{ii}^{(t)} + \Big\{\sum_{t=0}^{k} p_{ii}^{(t)}\, \ell_{ii}^{(k-t)}\Big\}\, p_{ij},$$

the first part following from the definition of the $\ell_{ij}^{(k)}$, in (5.1):

$$= \sum_{t=0}^{k-1} p_{ii}^{(t)}\, \ell_{ij}^{(k+1-t)} + p_{ii}^{(k)}\, p_{ij}$$

by the induction hypothesis;

$$= \sum_{t=0}^{k} p_{ii}^{(t)}\, \ell_{ij}^{(k+1-t)}$$

since $p_{ij} = \ell_{ij}$;

$$= \sum_{t=0}^{k+1} p_{ii}^{(t)}\, \ell_{ij}^{(k+1-t)}$$

since $\ell_{ij}^{(0)} = 0$. This completes the induction. #

The first set of relations between the generating functions now follows from the convolution structure of the relation just proved, if one bears in mind that $p_{ij}^{(0)} = \delta_{ij}$, by convention.

To prove the second part note that since $p_{ii}^{(0)} = 1$, for real z,

$$0 \leqslant z < 1, \quad 1 \leqslant P_{ii}(z) < \infty.$$

† This relation is a 'last exit' decomposition in relation to passage from i to j in an MC context.

Hence, considering such *real* z, we have

$$1 > 1 - \frac{1}{P_{ii}(z)} = L_{ii}(z).$$

Hence letting $z \to 1-$, $\quad 1 \geqslant L_{ii}(1-) = \sum_{k=0}^{\infty} \ell_{ii}^{(k)}$.

Thus for complex z satisfying $|z| < 1$,

$$|L_{ii}(z)| \leqslant L_{ii}(|z|) < 1.$$

Let us now define a (possibly infinite) quantity μ_i† by the limiting derivative

$$\mu_i = L'_{ii}(1-) \equiv \sum_{k=1}^{\infty} k\ell_{ii}^{(k)} \leqslant \infty,$$

for each index i.

DEFINITION 5.1. An index i is called *recurrent* if $L_{ii}(1-) = 1$, and *transient* if $L_{ii}(1-) < 1$.

A *recurrent* index i is said to be *positive- or null-recurrent* depending as $\mu_i < \infty$ or $\mu_i = \infty$ respectively. We call such μ_i the mean recurrence measure of a recurrent index i.

The following lemma and its corollary now establish a measure of relation between the old and new terminologies.

LEMMA 5.2. An inessential index is transient.

Proof. Let i be an inessential index; if i is not a member of a self-communicating class, i.e. $i \nrightarrow i$ then it follows from the definition of the $\ell_{ij}^{(k)}$ that $\ell_{ii}^{(k)} = 0$ all k, so that $L_{ii}(1) = 0 < 1$ as required.

Suppose now i is essential and a member of a self-communicating class, I. Then clearly (since $i \to i$) $\ell_{ii}^{(k)} > 0$ for some $k \geqslant 1$. This follows from the definition of the $\ell_{ij}^{(k)}$'s once more; as does the fact that such an $\ell_{ii}^{(k)}$ must consist of all non-zero summands of the form

$$p_{ir_1}\, p_{r_1 r_2} \cdots p_{r_{k-1} i}$$

where $r_j \neq i$, $j = 1, 2, \ldots, k-1$ if $k > 1$, $r_j \to i$; and simply of

$$p_{ii}$$

if $k = 1$. A non-zero summand of this form cannot involve an index u such that $i \to u$, $u \nrightarrow i$.

Now in I there must be an index i' such that there exists a $j \notin I$ satisfying $i \to j$, $j \nrightarrow i$, and $p_{i'j} > 0$, since i is inessential. It follows that there will be an index $q \in I$ such that

$$p_{i'q} > 0.$$

† The 'mean recurrence time' of state i in the MC context.

It follows moreover that this element $p_{i'q}$ will be one of the factors, for some k, of one of the non-zero summands mentioned above.

Now consider what happens if the matrix P is replaced by a new matrix $\hat{P} = \{\hat{p}_{rs}\}$ by altering the i'th row (only) of P, by way of putting $\hat{p}_{i'j} = 0$ for the i', j mentioned above, and scaling the other non-zero entries so that they still sum to unity. It follows in particular that $\hat{p}_{i'q} > p_{i'q}$.

It is then easily seen that for some $k \geqslant 1$

$$\hat{\ell}_{ii}^{(k)} > \ell_{ii}^{(k)} > 0$$

and that consequently

$$(1 \geqslant)\ \hat{L}_{ii}(1-) > L_{ii}(1-)$$

which yields the required result.† #

COROLLARY. A recurrent index is essential.

The following lemma provides an alternative criterion for distinguishing between the recurrence and transience of an index:

LEMMA 5.3. An index j is transient if and only if

$$\sum_{k=0}^{\infty} p_{jj}^{(k)} < \infty.$$

Proof. From Lemma 5.1, for real s, $0 \leqslant s < 1$

$$P_{jj}(s) = [1 - L_{jj}(s)]^{-1}$$

Letting $s \to 1-$ yields $P_{jj}(1-) \equiv \sum_{k=0}^{\infty} p_{jj}^{(k)} < \infty$, since $L_{jj}(1-) < \infty$. #

COROLLARY. An index j is recurrent if and only if

$$\sum_{k=0}^{\infty} p_{jj}^{(k)} = \infty.$$

In fact it is also true that if j is transient, then

$$\sum_{k=0}^{\infty} p_{ji}^{(k)} < \infty$$

for every index i such that $i \to j$, so that $p_{ji}^{(k)} \to 0$ as $k \to \infty$; but before we can prove this, we need to establish further relations between the generating functions $P_{ij}(z)$, and between the generating functions $L_{ij}(z)$.

LEMMA 5.4. For $|z| < 1$, and all $i, j = 1, 2, \ldots$

$$P_{ij}(z) = z \sum_r P_{ir}(z) p_{rj} + \delta_{ij}$$

$$L_{ij}(z) = z \sum_r L_{ir}(z) p_{rj} + z p_{ij}\,(1 - L_{ii}(z)).$$

† Which is, incidentally, trivial to prove by probabilistic, rather than the present analytical, reasoning.

Proof. The first relation follows directly from an elementwise consideration of the relation $P^{k+1} = P^k P$; and the second from substitution for the $P_{ij}(z)$ in terms of the $L_{ij}(z)$ via Lemma 5.1. #

COROLLARY 1. For each pair of indices i, j, such that $i \to j$, $L_{ji}(1-) < \infty$.

Proof. For s such that $0 < s < 1$, and all i, j,

$$s \sum_r L_{ir}(s) p_{rj} \leqslant L_{ij}(s)$$

from the second of the relations in Lemma 5.4 (noting $L_{ii}(s) < 1$). Iterating the above inequality k times,

$$s^k \sum_r L_{ir}(s) p_{rj}^{(k)} \leqslant L_{ij}(s), \text{ all } i, j, k \geqslant 1;$$

in particular

$$s^k \, L_{ir}(s) p_{rj}^{(k)} \leqslant L_{ij}(s), \text{ all } i, j, r, k \geqslant 1.$$

Thus, with the appropriate substitutions

$$s^k \, L_{ji}(s) p_{ij}^{(k)} \leqslant L_{jj}(s). \tag{5.3}$$

Now if $i \to j$, but $j \nrightarrow i$, the assertion of the Corollary 1 is trivial, so suppose that $i \to j$ and $j \to i$. Then there is a k such that $p_{ij}^{(k)} > 0$; and also from its definition, $L_{ji}(s) > 0$. Letting $s \to 1-$ in (5.3) yields $L_{ji}(1-) < \infty$ since $L_{jj}(1-) \leqslant 1$. Thus $0 < L_{ji}(1-) < \infty$. Thus in this case the Corollary is also valid. #

COROLLARY 2. If j is transient, and $i \to j$,

$$\sum_{k=0}^{\infty} p_{ji}^{(k)} < \infty$$

(and so $p_{ji}^{(k)} \to 0$ as $k \to \infty$). If j is recurrent, and $i \leftrightarrow j$,

$$\sum_{k=0}^{\infty} p_{ji}^{(k)} = \infty.$$

Proof. For $i = j$, this is covered by Lemma 5.3 and its Corollary. For $0 < s < 1$, and $i \neq j$, from Lemma 5.1

$$P_{ji}(s) = P_{jj}(s) L_{ji}(s).$$

Let $s \to 1-$.

Consider j transient first: then $P_{jj}(1-) < \infty$ by Lemma 5.3, and $L_{ji}(1-) < \infty$ by Corollary 1 above, so the result follows.

If j is recurrent, $P_{jj}(1-) = \infty$, and since $i \leftrightarrow j$, $0 < L_{ji}(1-), < \infty$; so that $P_{ji}(1-) = \infty$ as required. #

The problem of limiting behaviour for the $p_{ji}^{(k)}$ as $k \to \infty$ for recurrent indices is a deeper one, and we treat it in a separate section.

5.2. Limiting behaviour for recurrent indices

THEOREM 5.1. Let i be a recurrent *aperiodic*† index, with mean recurrence measure $\mu_i \leqslant \infty$. Then as $k \to \infty$

$$p_{ii}^{(k)} \to \mu_i^{-1}.$$

We shall actually prove a rather more general analytical form of the theorem.

THEOREM 5.1′. Let $f_0 = 0, f_j \geqslant 0, j \geqslant 1$, with

$$\sum_{j=0}^{\infty} f_j = 1.$$

Assume that the g.c.d. of those j for which $f_j > 0$, is unity‡, and let the sequence $\{u_k\}$, $k \geqslant 0$, be defined by

$$u_0 = 1; u_k = \sum_{j=0}^{k} f_j u_{k-j},\ k \geqslant 1.$$

It follows (e.g. by induction) that $0 \leqslant u_k \leqslant 1, k \geqslant 0$. Then as $k \to \infty$

$$u_k \to \frac{1}{\sum_{j=1}^{\infty} j f_j}.$$

[Theorem 5.1 now follows by putting $l_{ii}^{(j)} = f_j$, $p_{ii}^{(k)} = u_k$ for $j, k \geqslant 0$, in view of the basic relation

$$p_{ii}^{(k)} = \sum_{j=0}^{k} p_{ii}^{(j)} l_{ii}^{(k-j)} \equiv \sum_{j=0}^{k} l_{ii}^{(j)} p_{ii}^{(k-j)},\ k \geqslant 1$$

deduced in the proof of Lemma 5.1, in conjunction with our convention $p_{ii}^{(0)} = 1$.]

Proof. (i) We first note from Appendix A, that of those j for which $f_j > 0$, there is a finite subset $\{j_1, j_2, \ldots, j_r\}$ such that 1 is its g.c.d., and by the Corollary to Lemma A.3 any integer $q \geqslant N_0$, for some N_0 is expressible in the form

$$q = \sum_{t=1}^{r} p_t j_t,$$

with the p_t non-negative integers, depending on the value of q.

(ii) Put $r_t = \sum_{j=t}^{\infty} f_{j+1}$, noting $r_0 = 1$; then for $k \geqslant 1$,

$$r_0 u_k = \sum_{j=0}^{k} f_j u_{k-j} = \sum_{j=0}^{k-1} (r_j - r_{j+1}) u_{k-j-1},$$

† See Exercises 5.1 and 5.2 for an indication of the case of a periodic index.
‡ See Appendix A.

$$r_0 u_k + \sum_{j=0}^{k-1} r_{j+1} u_{k-(j+1)} \equiv \sum_{j=0}^{k} r_j u_{k-j} = \sum_{j=0}^{k-1} r_j u_{(k-1)-j}.$$

Thus for all $k \geqslant 1$,

$$\sum_{j=0}^{k} r_j u_{k-j} = r_0 u_0 = 1.$$

(iii) Since $0 \leqslant u_k \leqslant 1$, there exists a subsequence $\{k_\nu\}$ of the positive integers such that

$$\lim_{\nu \to \infty} u_{k_\nu - j} = \alpha \equiv \limsup_{k \to \infty} u_k \leqslant 1.$$

Let j be such that $f_j > 0$, and suppose

$$\limsup_{\nu \to \infty} u_{k_\nu - j} = \beta, \; (0 \leqslant \beta \leqslant \alpha).$$

Suppose $\alpha > \beta$. Then

$$u_{k_\nu} = \sum_{r=0}^{k_\nu} f_r u_{k_\nu - r}$$

$$\leqslant \sum_{\substack{r=0 \\ r \neq j}}^{M} f_r u_{k_\nu - r} + f_j u_{k_\nu - j} + \epsilon$$

where $0 < \epsilon < (\alpha - \beta) f_j$ and $M \geqslant j$ is chosen so that

$$r_M = \sum_{r=M+1}^{\infty} f_r < \epsilon,$$

since $0 \leqslant u_k \leqslant 1$. Letting $\nu \to \infty$

$$\alpha \leqslant \alpha \sum_{\substack{r=0 \\ r \neq j}}^{M} f_r + f_j \beta + \epsilon$$

i.e. letting $M \to \infty$,

$$\alpha \leqslant \alpha \sum_{\substack{r=0 \\ r \neq j}}^{\infty} f_r + f_j \beta + \epsilon$$

$$= \alpha \sum_{r=0}^{\infty} f_r - (\alpha - \beta) f_j + \epsilon$$

i.e. $$\alpha \leqslant \alpha - (\alpha - \beta) f_j + \epsilon.$$

Taking into account the choice of ϵ, this is a contradiction.
Hence $\beta = \alpha$; thus we may take $\{k_\nu\}$ to be a sequence such that

$$\lim_{\nu \to \infty} u_{k_\nu - j} = \alpha.$$

This is true for any j such that $f_j > 0$, the sequence $\{k_\nu\}$ depending on j. If we consider the set of j, $\{j_1, \ldots, j_r\}$ specified in part (i) of the proof, then by a suitable number of repetitions of the argument just given, for any fixed non-negative integers $p_1, p_2, \ldots, p_r$ there exists a (refined) sequence $\{k_\nu\}$ such that

$$\lim_{\nu\to\infty} u_{k_\nu-(p_1 j_1 + p_2 j_2 + \ldots + p_r j_r)} = \alpha$$

i.e. for any fixed integer $q \geqslant N_0$ (from part (i) of the proof)

$$\lim_{\nu\to\infty} u_{k_\nu - q} = \alpha \tag{5.4}$$

where subsequence $\{k_\nu\}$ depends on q.

Denote the sequence $\{k_\nu\}$ corresponding to $q = N_0$ by $\{k_\nu^{(0)}\}$. Repeat again the argument of (iii) in starting from *this* sequence to obtain a subsequence of it $\{k_\nu^{(1)}\}$ for which (5.4) holds, for $q = 1$, etc. It follows that, from the Cantor diagonal selection principle, there is a subsequence of the integers (we shall still call it $\{k_\nu\}$) such that

$$\lim_{\nu\to\infty} u_{k_\nu - q} = \alpha$$

for *every* $q \geqslant N_0$.

Now let $\gamma = \liminf_{k\to\infty} u_k$ $(0 \leqslant \gamma \leqslant \alpha \leqslant 1)$.

By an argument analogous but 'dual' to the one above,† there exists a subsequence $\{n_\nu\}$ of the integers such that

$$\lim_{\nu\to\infty} u_{n_\nu - q} = \gamma$$

for all $q \geqslant N_0$.

(iv) Now introduce new subsequences of the positive integers, $\{s_\nu\}, \{t_\nu\}$ defined for sufficiently large ν by

$$s_\nu = k_\nu - N_0, \quad t_\nu = n_\nu - N_0.$$

Then

$$\lim_{\nu\to\infty} u_{s_\nu - p} = \lim_{\nu\to\infty} u_{k_\nu-(N_0+p)} = \alpha$$

for $p \geqslant 0$, and similarly

$$\lim_{\nu\to\infty} u_{t_\nu - p} = \gamma.$$

Now

$$1 = \sum_{p=0}^{s_\nu} r_p u_{s_\nu - p},$$

$$1 = \sum_{p=0}^{t_\nu} r_p u_{t_\nu - p} \leqslant \sum_{p=0}^{M} r_p u_{t_\nu - p} + \epsilon$$

† But effectively using Fatou's Lemma rather than the epsilon argument.

where $\epsilon > 0$ is arbitrary, and M such that $r_M < \epsilon$. From these relations, as $\nu \to \infty$, taking lim inf and using Fatou's lemma in the first; and lim sup in the second, and subsequently $M \to \infty$; and $\epsilon \to 0$:

$$1 \geqslant \alpha \sum_{p=0}^{\infty} r_p, \quad 1 \leqslant \gamma \sum_{p=0}^{\infty} r_p$$

where $0 \leqslant \gamma \leqslant \alpha$.

Further

$$\sum_{p=0}^{\infty} r_p = \sum_{p=0}^{\infty} \sum_{j=p}^{\infty} f_{j+1} = \sum_{j=0}^{\infty} \sum_{p=0}^{j} f_{j+1} = \sum_{j=0}^{\infty} (j+1) f_{j+1}, = \mu, \text{ say.}$$

Thus if $\mu \equiv \sum_{p=0}^{\infty} r_p = \infty, \Rightarrow \alpha = 0$;

thus $\alpha = \gamma = \mu^{-1}$, as required.

If $$\mu < \infty, \mu^{-1} \geqslant \alpha \geqslant \gamma \geqslant \mu^{-1};$$

thus $\alpha = \gamma = \mu^{-1}$, as required.

This completes the proof. #

COROLLARY 1. If i is a recurrent aperiodic index and j is any index such that $j \to i$ then as $k \to \infty$

$$p_{ij}^{(k)} \to \mu_i^{-1} L_{ij}\,(1-).$$

Proof.

$$p_{ij}^{(k)} = \sum_{t=0}^{k} p_{ii}^{(t)}\, \ell_{ij}^{(k-t)} = \sum_{t=0}^{k} p_{ii}^{(k-t)}\, \ell_{ij}^{(t)}.$$

Now since $$L_{ij}(1-) = \sum_{k=0}^{\infty} \ell_{ij}^{(k)} < \infty$$

(by Corollary 1 to Lemma 5.4), we may use the Dominated Convergence Theorem to conclude that as $k \to \infty$ (by Theorem 5.1)

$$p_{ij}^{(k)} \to \mu_i^{-1} \sum_{t=0}^{\infty} \ell_{ij}^{(t)}$$

which is the required conclusion. #

COROLLARY 2. If i is in fact positive recurrent,

$$\sum_j L_{ij}(1-) \leqslant \mu_i < \infty.$$

Proof. By stochasticity and since a recurrent index is essential

$$1 = \sum_j p_{ij}^{(k)} = \sum_{\substack{j \\ j \to i}} p_{ij}^{(k)} + \sum_{\substack{j \\ j \not\to i}} p_{ij}^{(k)} = \sum_{\substack{j \\ j \to i}} p_{ij}^{(k)}$$

so that by Fatou's Lemma

$$1 \geqslant \mu_i^{-1} \sum_j L_{ij}(1-) \text{ as required.} \quad \#$$

[Note that by similar argument, Corollary 1 is trivially true if $j \nrightarrow i$]

It is possible to develop (as in the case of a finite matrix) a rather more unified theory in the important case of a single essential class of indices, i.e. in the case of an irreducible P. We now pass to this case.

5.3 Irreducible stochastic matrices

Since Lemma 1.2 of Chapter 1 continues to hold in the present context, we know that every index of an irreducible countable P has the *same* period, d.

Properties, like this, possessed in common by all indices of a single essential class of indices, are often called *solidarity* properties. We shall go on to show that transience, null-recurrence, and positive-recurrence are all solidarity properties, and to describe the limiting behaviour as $k \to \infty$ of P^k as a whole.

Moreover an important role in the present theory, is played by *subinvariant row vectors* (measures), and we treat this topic in some depth also. The results which are given for these are the (one-sided) analogues, for countable irreducible *stochastic P*, of the crucial Subinvariance Theorem of Chapter 1 for finite irreducible non-negative *T*.

THEOREM 5.2. All indices corresponding to an irreducible stochastic P are transient, or all are null-recurrent, or all are positive-recurrent.

Proof. We first note the elementary inequalities for any two indices r and j

$$p_{jj}^{(k+N+M)} \geqslant p_{jr}^{(N)} p_{rr}^{(k)} p_{rj}^{(M)}$$

$$p_{rr}^{(k+N+M)} \geqslant p_{rj}^{(M)} p_{jj}^{(k)} p_{jr}^{(N)}$$

which follow directly from $P^{k+N+M} = P^N P^k P^M = P^M P^k P^N$. Now since $r \leftrightarrow j$, it follows that M and N can be chosen so that

$$p_{rj}^{(M)} > 0, \, p_{jr}^{(N)} > 0.$$

Now the index r must be transient, or null-recurrent, or positive-recurrent. We treat the cases separately.

(i) If r is transient

$$\sum_{k=0}^{\infty} p_{rr}^{(k)} < \infty$$

from Lemma 5.3; this implies by the second of the elementary inequalities that

$$\sum_{k=0}^{\infty} p_{jj}^{(k)} < \infty$$

so (any other index) j is also transient.

(ii) If r is null-recurrent, from Theorem 5.1

$$p_{rr}^{(k)} \to 0 \text{ as } k \to \infty;$$

thus by the second of the elementary inequalities

$$p_{jj}^{(k)} \to 0 \text{ as } k \to \infty.$$

On the other hand, since r is recurrent, from the Corollary to Lemma 5.3 it follows that

$$\sum_{k=0}^{\infty} p_{rr}^{(k)} = \infty;$$

hence the first of the elementary inequalities implies

$$\sum_{k=0}^{\infty} p_{jj}^{(k)} = \infty$$

so that by Lemma 5.3, j is recurrent. Now if j is assumed positive recurrent, we know from Theorem 5.1 (and its periodic version) that as $k \to \infty$

$$p_{jj}^{(k)} \nrightarrow 0;$$

which is a contradiction. Hence j is also null-recurrent.

(iii) If r is positive recurrent, j must be also; otherwise a contradiction to the positive recurrence of r would arise from (i) or (ii). #

This theorem justifies the following definition.

DEFINITION 5.2. An irreducible P is said to be transient, or null-recurrent, or positive-recurrent depending on whether any one of its indices is transient, or null-recurrent, or positive-recurrent, respectively.

DEFINITION 5.3. For a stochastic P, a row vector $\mathbf{x}'$, $\mathbf{x}' \geqslant \mathbf{0}'$, $\mathbf{x}' \neq \mathbf{0}'$ satisfying

$$\mathbf{x}'P \leqslant \mathbf{x}'$$

is called a *subinvariant* measure. If in fact $\mathbf{x}'P = \mathbf{x}'$ the $\mathbf{x}'$ is called an *invariant* measure. [Note that a positive multiple of such a measure is still such a measure.]

LEMMA 5.5. For a stochastic irreducible P, a subinvariant measure always exists. (One such is given by the vector $\{L_{ij}(1-)\}$, $j = 1, 2,$ [for arbitrary fixed i], whose elements are positive and finite.)

Proof. From Lemma 5.4, for i fixed but arbitrary, and $j = 1, 2, \ldots$, and $0 < s < 1$

$$L_{ij}(s) = s\ \Sigma L_{ir}(s)\, p_{rj} + s\, p_{ij}(1 - L_{ii}(s)). \tag{5.5}$$

Letting $s \to 1-$ and using Fatou's Lemma

$$L_{ij}(1-) \geqslant \sum_r L_{ir}(1-)\, p_{rj} + p_{ij}(1 - L_{ii}(1-))$$

whence, since $L_{ii}(1-) \leqslant 1$, the row vector $\{L_{ij}(1-)\}$, $j = 1, 2, \ldots$ is a subinvariant measure. We know that since $i \leftrightarrow j$

$$0 < L_{ij}(1-) < \infty$$

from Corollary 1 of Lemma 5.4.

COROLLARY For fixed $i, j = 1, 2, 3, \ldots$

$$L_{ij}(1-) = \sum_r L_{ir}(1-)\, p_{rj} + p_{ij}(1 - L_{ii}(1-)).$$

Proof. From the proof above, since

$$s \sum_r L_{ir}(s)\, p_{rj} \leqslant \sum_r L_{ir}(1-)\, p_{rj} \leqslant L_{ij}(1-) < \infty$$

we may use the Dominated Convergence Theorem in (5.5) in letting $s \to 1-$, to obtain the 'finer' result stated.

LEMMA 5.6. Any subinvariant measure for irreducible stochastic P has all its entries positive.

Proof. For such a measure $\boldsymbol{x}'$

$$\boldsymbol{x}'P \leqslant \boldsymbol{x}'$$

so that

$$\boldsymbol{x}'P^k \leqslant \boldsymbol{x}'$$

and in particular

$$\sum_i x_i p_{ij}^{(k)} \leqslant x_j$$

i.e. for any i, j

$$x_i p_{ij}^{(k)} \leqslant x_j.$$

Select fixed i so that $x_i > 0$; for any fixed j, since $i \to j$, there is a k such that $p_{ij}^{(k)} > 0$. Hence

$$x_j > 0 \quad \text{all } j = 1, 2, \ldots \quad \#$$

Lemma 5.6 makes sensible the formulation of the following result.

THEOREM 5.3. If $\boldsymbol{x}' = \{x_i\}$ is any subinvariant measure corresponding to irreducible stochastic P, then for fixed but arbitrary i, and all $j = 1, 2, \ldots$

$$x_j / x_i \geqslant \bar{x}_{ij}$$

where†

$$\bar{x}_{ij} = (1 - \delta_{ij})\, L_{ij}(1-) + \delta_{ij}$$

and $\bar{x}_{ij}$, $j = 1, 2, \ldots$ is also a subinvariant measure (with ith element unity).

† δ_{ij} is the Kronecker delta, as before.

[This theorem therefore says that out of all subinvariant measures normed to have a fixed element unity, there is one which is *minimal.*]

Proof. That $\{\bar{x}_{ij}\}, j = 1, 2, \ldots$ is subinvariant is readily checked from the statement of Corollary 1 of Lemma 5.5, since $L_{ii}(1-) \leqslant 1$.

We prove the rest by induction. Let $\{y_j\}$ be any subinvariant measure with $y_i = 1$. It is required to show that for every j

$$y_j \geqslant (1 - \delta_{ij}) \sum_{k=0}^{\infty} \ell_{ij}^{(k)} + \delta_{ij}$$

or equivalently that for every j, and every $m \geqslant 1$

$$y_j \geqslant (1 - \delta_{ij}) \sum_{k=0}^{m} \ell_{ij}^{(k)} + \delta_{ij}.$$

Now we have for all j,

$$y_j \geqslant \sum_r y_r p_{rj} \geqslant y_i p_{ij} = \ell_{ij}^{(1)}$$

and by assumption

$$y_i = 1,$$

so the proposition is true for $m = 1$ and all j. Assume it is true for $m \geqslant 1$; then for each j

$$\begin{aligned} y_j &\geqslant \sum_r y_r p_{rj} \\ &= \sum_{r \neq i} y_r p_{rj} + p_{ij} \\ &= \sum_{r \neq i} \sum_{k=0}^{m} \ell_{ir}^{(k)} p_{rj} + p_{ij} \\ &= \sum_{k=1}^{m} \sum_{r \neq i} \ell_{ir}^{(k)} p_{rj} + p_{ij} \end{aligned}$$

and from the definition of the $\ell_{ij}^{(k)}$

$$\begin{aligned} &= \sum_{k=1}^{m} \ell_{ij}^{(k+1)} + \ell_{ij}^{(1)} \\ &= \sum_{k=0}^{m} \ell_{ij}^{(k+1)} = \sum_{k=0}^{m+1} \ell_{ij}^{(k)}. \end{aligned}$$

Hence
$$y_j \geqslant \sum_{k=0}^{m+1} \ell_{ij}^{(k)}, \text{ all } j,$$

and moreover

$$y_i = 1$$

by given, so induction is true for $m + 1$; This completes the proof. #

THEOREM 5.4. For a recurrent† matrix P an invariant measure exists, and is unique (to constant multiples).

A subinvariant measure which is not invariant exists if and only if P is a transient matrix; one such subinvariant measure is then given by $\{\bar{x}_{ij}\}$, $j = 1, 2, \ldots$

Proof. If P is recurrent, $L_{ii}(1-) = 1$ for all i, so existence of an invariant measure $\{\bar{x}_{ij}\}$, $j = 1, 2, \ldots$ follows from the Corollary of Lemma 5.5. Now let $\boldsymbol{y} = \{y_j\}$ be any subinvariant measure scaled so that $y_i = 1$ for fixed i. Then

$$\boldsymbol{z} = \{y_j - \bar{x}_{ij}\}, j = 1, 2, \ldots$$

satisfies $\boldsymbol{z} \geqslant \boldsymbol{0}$, (from Theorem 5.3) and $z_i = 0$; and if $\boldsymbol{z} \neq \boldsymbol{0}$, $\boldsymbol{z}$ is clearly a subinvariant measure, since $\boldsymbol{y}$ is subinvariant and $\{\bar{x}_{ij}\}$ is invariant: but $z_i = 0$ and this is not possible by Lemma 5.6.

Hence $\boldsymbol{z} = \boldsymbol{0}$, which completes the proof for the recurrent case.

If P is transient, $L_{ii}(1-) < 1$ for each i, hence a subinvariant measure which is not invariant may be taken as $\{\bar{x}_{ij}\}$, $j = 1, 2, \ldots$ for fixed i from the Corollary of Lemma 5.5; there is strict inequality in the subinvariance equations only at the ith position. Suppose now a subinvariant measure $\boldsymbol{y} = \{y_j\}$, normed so that $y_i = 1$, exists but is not invariant, in relation to a stochastic P. Then

$$y_j \geqslant \bar{x}_{ij}, j = 1, 2, \ldots$$

If P is recurrent, then from the first part $\boldsymbol{y}$ must be invariant, since $y_j = \bar{x}_{ij}$, which is a contradicition. Hence P must be transient. #

COROLLARY If $\boldsymbol{x}'$ is a subinvariant measure for recurrent P, it is, in fact, an invariant measure, and is a positive multiple of $\{\bar{x}_{ij}\}$, $j = 1, 2, \ldots$

THEOREM 5.5. (General Ergodic Theorem‡). Let P be a primitive ¶ stochastic matrix. If P is transient or null-recurrent, then for each pair of indices i, j, $p_{ij}^{(k)} \to 0$ as $k \to \infty$.

If P is positive recurrent, then for each pair i, j

$$\lim_{k \to \infty} p_{ij}^{(k)} = \mu_j^{-1}$$

and the vector $\boldsymbol{\nu} = \{\mu_j^{-1}\}$ is the unique invariant measure of P satisfying $\boldsymbol{\nu}'\mathbf{1} = 1$. [$\boldsymbol{\nu}$ is thus of course the unique stationary distribution.]

Proof. If P is transient, the result follows from Corollary 2 of Lemma 5.4 (in fact if P is merely irreducible); if P is null-recurrent, from Corollary 1 of Theorem 5.1.

† i.e. positive-recurrent or null-recurrent

‡ cf. Theorem 4.2.

¶ For the irreducible periodic case see Exercises 5.1 and 5.2.

If P is positive recurrent on the other hand, for any pair i, j

$$p_{ij}^{(k)} \to \mu_i^{-1} L_{ij}(1-) = \mu_i^{-1} \bar{x}_{ij}$$

(since in the recurrent case $L_{ii}(1-) = 1$).

From the Corollary to Theorem 5.4, for fixed i

$$\sum_r \bar{x}_{ir} p_{rj} = \bar{x}_{ij}, \; j = 1, 2, \ldots$$

where (from Corollary 2 of Theorem 5.1)

$$\sum_r \bar{x}_{ir} \leqslant \mu_i < \infty.$$

Hence $u' = \{u_j\}$ where

$$u_j = \bar{x}_{ij} / \sum_r \bar{x}_{ir}$$

is the unique invariant measure of P satisfying $u'\mathbf{1} = 1$ (and hence is independent of the initial choice of i).

Now since $u'P = u'$ it follows $u'P^k = u'$ so that

$$u_j = \sum_r u_r p_{rj}^{(k)}, \text{ all } j.$$

Since $\sum_r u_r < \infty$, we have, by dominated convergence, letting $k \to \infty$

$$u_j = \sum_r u_r \mu_r^{-1} \bar{x}_{rj}. \tag{5.6}$$

Now suppose that in fact

$$\sum_j \bar{x}_{rj} < \mu_r \text{ for some } r.$$

Then summing over j in (5.6) (using Fubini's Theorem)

$$\sum_j u_j < \sum_r u_r$$

which is impossible. Hence

$$\sum_j \bar{x}_{rj} = \mu_r \text{ for all } r = 1, 2, \ldots$$

and so

$$u_j = \mu_i^{-1} \bar{x}_{ij} \text{ for all } i;$$

and by putting $i = j$, we see

$$u_j = v_j = \mu_j^{-1}$$

as required, since $\bar{x}_{ii} = 1$. #

COROLLARY. If P is an irreducible transient or null-recurrent matrix, there exists no invariant measure $\boldsymbol{v}$ satisfying $\boldsymbol{v}'\mathbf{1} < \infty$.

Proof. In either case† we know $p_{ij}^{(k)} \to 0$ as $k \to \infty$. Suppose a measure of the required sort exists: then

$$v_j = \sum_i v_i p_{ij} = \sum_i v_i p_{ij}^{(k)}.$$

Since $\sum_i v < \infty$, by dominated convergence

$$v_j = \sum_i v_i (\lim_{k \to \infty} p_{ij}^{(k)}) = 0,$$

for each $j = 1, 2, \ldots$, which is a contradiction to $\boldsymbol{v} \geqslant \mathbf{0}, \neq \mathbf{0}$. #

5.4 The 'dual' approach; subinvariant vectors

In §§5.1–5.3 of this chapter we developed the theory of classification of indices, subinvariant measures, and asymptotic behaviour of powers P^k, for a countable stochastic matrix P, in terms of the quantities $\ell_{ij}^{(k)}$ and consequently the $L_{ij}(1-)$.

There is a more classical approach, dual to the one we have presented, for developing the index classification and behaviour of powers theory, which is in many respects (as will be seen) more natural than the one just given, since it accords better with the fact that the row sums of P are unity. However (although it too leads to Theorem 5.5), it is not suitable for dealing with subinvariant measures, which multiply P from the left, and are particularly important in the stability theory of Markov chains: but rather with *subinvariant vectors.*

DEFINITION 5.4. For a stochastic P, a column vector $\boldsymbol{u}$, $\boldsymbol{u} \geqslant \mathbf{0}$, $\boldsymbol{u} \neq \mathbf{0}$ satisfying

$$P\boldsymbol{u} \leqslant \boldsymbol{u}$$

is called a *subinvariant vector.* If in fact $P\boldsymbol{u} = \boldsymbol{u}$, the vector $\boldsymbol{u}$ is called an *invariant vector.* [A positive multiple of such is still such a vector.]

The stochasticity of P ensures that an invariant vector always exists, viz. $\mathbf{1}$ is such, for $P\mathbf{1} = \mathbf{1}$. While subinvariant vectors are of little importance in the stability theory of MC's, they are of substantial interest in the description of long term behaviour of transient chains, i.e. of their Martin exit boundary theory, to which we shall pass shortly.

It is thus appropriate to sketch here the alternative development, although all proofs will be left to the interested reader.‡

The development is in terms of the quantities $f_{ij}^{(k)}$ $k \geqslant 0, i, j = 1, 2, \ldots$ defined by

$$f_{ij}^{(1)} = p_{ij};\, f_{ij}^{(k+1)} = \sum_{\substack{r \\ r \neq j}} p_{ir} f_{rj}^{(k)},\; k \geqslant 1$$

where $f_{ij}^{(0)} = 0$ by definition. [In the MC framework, $f_{ij}^{(k)}$ is the probability of

† See Exercise 5.2 for the periodic situation.

‡ See Exercise 5.4.

going from i to j in k steps, without visiting j in the meantime; it is also an example of a 'taboo' probability, and the set $\{f_{ij}^{(k)}\}$, $k = 0, 1, 2, \ldots$ is known as the first passage time distribution from i to j.]

The generating functions, well defined $|z| < 1$,

$$F_{ij}(z) = \sum_k f_{ij}^{(k)} z^k, \; P_{ij}(z) = \sum_k p_{ij}^{(k)} z^k$$

are related by (the analogue of Lemma 5.1)

$$P_{ii}(z) = [1 - F_{ii}(z)]^{-1}; P_{ij}(z) = F_{ij}(z)\,P_{jj}(z), i \neq j$$

from which it is seen that $F_{ii}(z) = L_{ii}(z)$, $|z| < 1$, so that $f_{ii}^{(k)} = \ell_{ii}^{(k)}$. From this it is seen that the classification theory of indices is the same, and Theorem 5.1 is equally relevant to the present situation.

On the other hand, instead of Lemma 5.4 we have from considering now $P^{k+1} = PP^k$:

LEMMA 5.4.^{D}. For $|z| < 1$, and all $i, j = 1, 2, \ldots$

$$P_{ij}(z) = z \sum_r p_{ir} P_{rj}(z) + \delta_{ij}$$

$$F_{ij}(z) = z \sum_r p_{ir} F_{rj}(z) + z\, p_{ij}\,(1 - F_{jj}(z)).$$

COROLLARY 1. For each pair of indices i, j, $F_{ij}(1-) \leq 1$.

COROLLARY 2. If j is transient,

$$\sum_{k=0}^{\infty} p_{ij}^{(k)} < \infty$$

for all i (and so $p_{ij}^{(k)} \to 0$ as $k \to \infty$). If i is recurrent and $i \leftrightarrow j$,

$$\sum_{k=0}^{\infty} p_{ij}^{(k)} = \infty.$$

Instead of Corollary 1 to Theorem 5.1, we have that if j is a *recurrent aperiodic index*, and i is any other index,

$$p_{ij}^{(k)} \to \mu_j^{-1} F_{ij}\,(1-).$$

The results are adequate to prove the analogue of Theorem 5.2. The analogue of Lemma 5.5 is trivial in view of the vector $\boldsymbol{u} = \mathbf{1}$; and the analogue of Lemma 5.6 holds from similar argument.

The minimal subinvariant vector is given, for fixed but arbitrary j and $i = 1, 2, \ldots$ by

$$\bar{u}_{ij} = (1 - \delta_{ij})F_{ij}(1-) + \delta_{ij};$$

and the analogue of Theorem 5.4 reads:

THEOREM 5.4^{D}. For a recurrent matrix P, an invariant vector exists and is a positive multiple of the vector $\mathbf{1}$, and $\bar{u}_{ij} = 1$, all i, j.

A subinvariant vector which is not invariant exists if and only if P is a transient matrix; one such subinvariant vector is given by $\{\bar{u}_{ij}\}$, $i = 1, 2, \ldots$

Theorem 5.5 can be proved also, using slightly different emphasis. Some of the notes made in the present section will be required in the next.

5.5 Potential and boundary theory for transient indices†

In this section only we shall relax the assumption that $P = \{p_{ij}\}$, $i, j \geq 1$ is stochastic in a minor way, assuming only that

$$p_{ij} \geq 0,\ 1 \geq \sum_j p_{ij} > 0 \text{ each } i.$$

In the strictly substochastic case, i.e. where for some i, $1 > \sum_j p_{ij}$, we may think of P as a part of an enlarged stochastic matrix $\bar{P} = \{\bar{p}_{ij}\}$, on the index set $\{0, 1, 2, \ldots\}$, where $\bar{p}_{ij} = p_{ij}$, $i, j \geq 1$; $\bar{p}_{i0} = 1 - \sum_j p_{ij}$, $i \geq 1$, $\bar{p}_{00} = 1$: in this case

$$\bar{P} = \begin{bmatrix} 1 & \mathbf{0}' \\ \mathbf{p}_1 & P \end{bmatrix}, \mathbf{p}_1 \geq \mathbf{0}, \neq \mathbf{0}$$

so that

$$\bar{P}^k = \begin{bmatrix} 1 & \mathbf{0}' \\ \mathbf{p}_k & P^k \end{bmatrix};$$

so in either situation it is meaningful to study the (well-defined) powers P^k of the matrix $P = \{p_{ij}\}$, $i, j = 1, 2, \ldots$

It is readily seen that Lemma 5.4^D and its Corollaries continue to hold in the present more general situation; and in the strictly substochastic case of P, the $P_{ij}(z)$ and $F_{ij}(z)$ for $i, j \geq 1$ coincide with the corresponding quantities $\bar{P}_{ij}(z)$, $\bar{F}_{ij}(z)$ of the expanded matrix $\bar{P}$ for $i, j \geq 1$ ($\bar{f}_{0j}^{(k)} = 0$ for all $k \geq 0$ if $j \geq 1$).

We shall now extend slightly the notion of a subinvariant vector, and introduce the more usual terminology which generally occurs in this context.

DEFINITION 5.5. A column vector‡ $\mathbf{u} \geq \mathbf{0}$ satisfying

$$P\mathbf{u} \leq \mathbf{u}$$

is said to be a *superregular vector* for P. If in fact $P\mathbf{u} = \mathbf{u}$, $\mathbf{u}$ is said to be *regular.*

Thus the vector $\mathbf{0}$ is regular for P; and the vector $\mathbf{1}$ is superregular (but is not regular if P is strictly substochastic).

† This section is of more specialist interest; and in the part dealing with the Poisson–Martin representation requires deeper mathematics than the rest of the book.

‡ Assumed elementwise finite as usual.

If we define the vector $\{\bar{u}_{ij}\}$, $i = 1, 2, \ldots$ for fixed but arbitrary $j \geqslant 1$ as in the previous section, it will similarly be superregular for P: in fact

$$\bar{u}_{ij} = \sum_k p_{ik}\bar{u}_{kj} + \delta_{ij}\,(1 - F_{jj}(1-)),\; i = 1, 2, \ldots \tag{5.7}$$

[so that strict inequality may occur only in position $i = j$ as regards the superregularity equations]; and the minimality property in this case (obtained with an almost identical induction proof) will take the form that for any superregular $\boldsymbol{u}$,

$$u_i \geqslant \bar{u}_{ij}u_j,\; i, j = 1, 2, \ldots$$

(since we no longer have the assurance that $\boldsymbol{u} > \mathbf{0}$).

Since $\{\bar{u}_{ir}\}$, $i = 1, 2, \ldots$ is itself superregular, we have, by putting $u_i = \bar{u}_{ir}$, $i = 1, 2, \ldots$:

> The Fundamental Inequality: $\bar{u}_{ir} \geqslant \bar{u}_{ij}\bar{u}_{jr}$, all $i, j, r \geqslant 1$.

We shall for the rest of the section make the

Basic Assumption 1. (i) The index set of P is in fact *all the positive integers*, R. (The case of finite P is excluded here for the first time.)

(ii) All indices are *transient*: i.e. $F_{ii}(1-) < 1$ for all $i \geqslant 1$.

[Note that we do *not* necessarily have irreducibility of P.]

By Corollary 2 of Lemma 5.4^{D} of the previous section

$$g_{ij} = \sum_{k=0}^{\infty} p_{ij}^{(k)} < \infty \quad \text{for all } i, j \in R$$

so that the non-negative matrix $G = \{g_{ij}\}$ on $i, j \in R$ (i.e. $R \times R$) is elementwise finite. This matrix will be used to define potentials on R: it is sometimes called the *Green's function*, or *kernel* defined on $(R \times R)$ corresponding to P.

Potential theory

Since $\boldsymbol{u} \geqslant P\boldsymbol{u}$ for a vector $\boldsymbol{u}$ superregular on R, it follows

$$\boldsymbol{u} \geqslant P\boldsymbol{u} \geqslant P^2\boldsymbol{u} \geqslant \ldots \geqslant P^k\boldsymbol{u} \geqslant \mathbf{0}.$$

Denote the limit vector of the monotone decreasing non-negative vector sequence $\{P^k\boldsymbol{u}\}$ by $P^{\infty}\boldsymbol{u}$.

LEMMA 5.7. For a superregular vector $\boldsymbol{u}$, $P^{\infty}\boldsymbol{u}$ is regular on R, and moreover

$$\boldsymbol{u} = P^{\infty}\boldsymbol{u} + G(\boldsymbol{u} - P\boldsymbol{u}).$$

Proof.

$$P^{\infty}\boldsymbol{u} = \lim_{k\to\infty} P^{r+k}\boldsymbol{u} = \lim_{k\to\infty} P^r\,(P^k\boldsymbol{u})$$

(by Fubini's Theorem), and since $P^k u \leqslant u$, and $P^k u \downarrow P^\infty u$,

$$= P^r (\lim_{k\to\infty} P^k u)$$

by dominated convergence,

$$\equiv P^r (P^\infty u).$$

Taking $r = 1$ shows that $P^\infty u$ is regular.
Now, we may write u as

$$u = P^{r+1} u + \sum_{k=0}^{r} (P^k u - P^{k+1} u), \text{ for } r \geqslant 1$$

$$= P^{r+1} u + \sum_{k=0}^{r} P^k [u - Pu].$$

Letting $r \to \infty$ yields the result, since $G = \sum_{k=0}^{\infty} P^k$. #

DEFINITION 5.6. If $\nu \geqslant \mathbf{0}$ (elementwise finite as usual) is a column vector, then $G\nu$ is called its potential [$G\nu$ may have some entries infinite.]

LEMMA 5.8. If a potential $G\nu$ is everywhere finite, it (*a*) determines ν; and, (*b*) is superregular.

Proof. It is easily checked that

$$G = PG + I$$

from the definition of G; applying this to (an elementwise finite) vector ν we have

$$G\nu = P(G\nu) + \nu.$$

Thus if $G\nu$ is everywhere finite, $P(G\nu)$ is finite, and

$$\nu = G\nu - P(G\nu) = (I - P)(G\nu)$$

which proves the first part; further in this situation clearly

$$G\nu \geqslant P(G\nu)$$

which is equivalent to the second part. #

LEMMA 5.9. A necessary and sufficient condition for a superregular vector u to be a (finite) potential is $P^\infty u = \mathbf{0}$.

Proof. Suppose a non-negative vector u is an elementwise finite potential; then for some (elementwise finite) $\nu \geqslant 0$

$$u = G\nu,$$

$$= Pu + \nu$$

from the proof of Lemma 5.8, i.e.

$$P^k u - P^{k+1} u = P^k v$$

i.e.
$$\sum_{k=0}^{r} (P^k u - P^{k+1} u) = \sum_{k=0}^{r} P^k v$$

$$u - P^{r+1} u = \left(\sum_{k=0}^{r} P^k\right) v.$$

Let $r \to \infty$;

$$u - P^{\infty} u = Gv, = u \text{ by definition of } u \text{ and } v.$$

Therefore
$$P^{\infty} u = \mathbf{0}.$$

Now let u be a superregular vector such that $P^{\infty} u = \mathbf{0}$. Define v, non-negative and elementwise finite by

$$v = u - Pu.$$

Hence as before

$$\sum_{k=0}^{r} P^k v = \sum_{k=0}^{r} (P^k u - P^{k+1} u)$$

and letting $r \to \infty$

$$Gv = u - P^{\infty} u$$

$$= u$$

by assumption. Hence u is a potential for v. #

THEOREM 5.6. A superregular vector u may be decomposed into a sum

$$u = r + g$$

where g is a potential and r is regular. The decomposition is unique; in fact

$$r = P^{\infty} u, \; g = G(u - Pu).$$

Proof. Lemma 5.7 already asserts the possibility of decomposition with the specific forms for r and g stated. To prove uniqueness let

$$u = r + g$$

where r is regular and g is a potential; g is necessarily elementwise finite, since u is, by its very definition.

Then

$$P^k u = P^k r + P^k g$$

i.e.
$$P^k u = r + P^k g$$

since r is regular; let $k \to \infty$, taking into account g is also superregular, in view of Lemma 5.8

$$P^{\infty} u = r + P^{\infty} g$$

$$= r$$

since $P^{\infty} g = \mathbf{0}$ by Lemma 5.9.

Hence

$$u = P^{\infty} u + g$$

and the fact that $g = G(u - Pu)$ follows from Lemma 5.7. #

THEOREM 5.7. Let u be a superregular vector and h an (elementwise finite) potential. Then (*a*) the elementwise minimum, $c = \min(u, h)$ is also a potential; and (*b*) there exists a non-decreasing sequence of (finite) potentials $\{h_n\}$ converging (elementwise) to u.

Proof. (*a*) since h is a finite potential, $P^{\infty} h = 0$ (Lemma 5.9).

Thus since $h \geqslant c \geqslant 0$, then $P^k h \geqslant P^k c \geqslant 0$.

It follows, letting $k \to \infty$, that $P^{\infty} c = 0$.

We shall now prove c is superregular (it is clearly elementwise finite).

$$Pc \leqslant Pg \leqslant h$$

since h, being a potential, is superregular; and from definition of c. Similarly

$$Pc \leqslant Pu \leqslant u$$

from definition of c and u. Hence

$$Pc \leqslant \min(u, h) = c.$$

Hence by Lemma 5.9, c must be a potential.

(*b*) Let g and h be any two finite potentials; then clearly $g + h$ is superregular (since both g and h are) and

$$P^{\infty} (g + h) = P^{\infty} g + P^{\infty} h = 0 + 0 = 0$$

by Lemma 5.9. Hence again by Lemma 5.9, $g + h$ is a finite potential; thus a sum of finite potentials is itself a finite potential.

Now let f_j be the column vector with unity in the jth position and zeros elsewhere: since $g_{jj} > 0$ always, Gf_j is a potential having its jth position non-zero, at least.

Let

$$g_j = x(j)Gf_j$$

where $x(j)$ is positive integer chosen so that $x(j)g_{jj} > u(j)$ where $u = \{u(j)\}$ is the given superregular vector. Then g_j is a finite potential (being a finite sum of finite potentials), and d_n defined by

$$d_n = \sum_{j=1}^{n} g_j$$

is also. It follows that $d_n(j) > u(j), j = 1, 2, \ldots, n$.

Consider now the sequence $\{h_n\}$ defined by

$$h_n = \min(u, d_n),$$

$\boldsymbol{h}_n$ being a potential from the first part (a) of the theorem. Moreover $\{\boldsymbol{h}_n\}$ is a non-decreasing sequence of finite potentials converging elementwise to $\boldsymbol{u}$. #

The reader acquainted with various versions of potential theory in physics will recognize that the above results are formally analogous. Theorem 5.6 is, for example, the analogue of the Riesz Decomposition Theorem; some further discussion will be given at the end of the chapter. We shall have occasion to make use of some of the previously developed results shortly.

The Martin exit boundary; the Poisson–Martin integral representation for a superregular vector

The basic purpose of this subsection, in keeping with the general aims of the book, is to develop an important representation of a superregular vector which is similar to, but rather more sophisticated than, the Riesz decomposition given by Lemma 5.7 and Theorem 5.6. In this connection we need to develop first a small amount of boundary theory, the Martin exit boundary being of additional importance in the study of long-term behaviour of the trajectories of infinite Markov chains which consist of transient states only. This probabilistic framework, which we shall not pursue here, nevertheless motivates the making of an additional basic assumption, which can in effect be reasoned (from the probabilistic framework) to be made without essential loss of generality.

Basic Assumption 2. The index 1 leads to every index, i.e. $1 \to j$, $j = 1, 2, \ldots$ [This implies that in $G = \{g_{ij}\}$, $g_{1j} > 0$ for all $j \in R$.]

Define for all $i, j \in R$

$$K(i, j) = \frac{\bar{u}_{ij}}{\bar{u}_{1j}} \text{ (where } \bar{u}_{1j} > 0 \text{ for all } j \in R \text{ since } 1 \to j)$$

the $\bar{u}_{ij} = (1 - \delta_{ij}) F_{ij}(1-) + \delta_{ij}$ having been defined in §5.4. [It may be useful to note also, on account of the relation between the $F_{ij}(z)$ and $P_{ij}(z)$, that in fact

$$K(i, j) = \frac{g_{ij}}{g_{1j}}, \text{ all } i, j \in R.]$$

From the Fundamental Inequality deduced earlier in this section

$$\bar{u}_{1j} \geqslant \bar{u}_{1i}\, \bar{u}_{ij}$$

so that

$$K(i, j) \leqslant 1/\bar{u}_{1i} \text{ for all } i, j;$$

and further, since for fixed $j \in R$, $\{\bar{u}_{ij}\}$ is superregular, then so is $K(i, j)$.

Let us now define on $R \times R$ the function

$$d(\nu_1, \nu_2) = \sum_{i \in R} |K(i, \nu_1) - K(i, \nu_2)| \bar{u}_{1i} w_i$$

where $\{w_i\}$ is any sequence of strictly positive numbers such that

$$\sum_{i \in R} w_i < \infty$$

(e.g. we might take $w_i = 2^{-i}$).

We shall now show that $d(\,.\,,\,.\,)$ is a *metric* on R i.e.

$$0 \leqslant d(\nu_1, \nu_2) < \infty \text{ for } \nu_1, \nu_2 \in R$$

and: (i) $d(\nu_1, \nu_2) = 0$ if and only if $\nu_1 = \nu_2$;
(ii) $d(\nu_1, \nu_2) = d(\nu_2, \nu_1)$;
(iii) $d(\nu_1, \nu_3) \leqslant d(\nu_1, \nu_2) + d(\nu_2, \nu_3)$.

The non-negativity is trivial; and finiteness of the bivariate function $d(\,.\,,\,.\,)$ follows from the fact that

$$d(\nu_1, \nu_2) \leqslant 2 \sum_{i \in R} w_i < \infty$$

on account of the triangle inequality on the real numbers, and the bound on $K(i, j)$ established above. Similarly (ii) and (iii) are trivial; and in fact the only non-obvious proposition which needs to be demonstrated is that

$$d(\nu_1, \nu_2) = 0 \Rightarrow \nu_1 = \nu_2 .$$

Suppose not; suppose that for some ν_1, ν_2, $d(\nu_1, \nu_2) = 0$ and $\nu_1 \neq \nu_2$. Now $d(\nu_1, \nu_2) = 0$ implies

$$K(i, \nu_1) = K(i, \nu_2), \quad \textit{all } i \in R$$

since $\bar{u}_{1i}\, w_i > 0$ for all i. Thus

$$\sum_{i \in R} p_{\nu_2 i} K(i, \nu_1) = \sum_{i \in R} p_{\nu_2 i} K(i, \nu_2).$$

But since $\nu_1 \neq \nu_2$ by assumption, this becomes

$$K(\nu_2, \nu_1) = K(\nu_2, \nu_2) - \frac{(1 - F_{\nu_2 \nu_2}(1-))}{\bar{u}_{1\nu_2}}$$

from the subinvariance equations (5.7) for $\{\bar{u}_{ij}\}$, $i = 1, 2, \ldots$ for fixed j, where inequality occurs only in the $i = j$ position.

i.e. $$K(\nu_2, \nu_1) < K(\nu_2, \nu_2)$$

which is a contradiction to $K(i, \nu_1) = K(i, \nu_2)$ for all $i \in R$.

Thus R is metrized by the metric d, and so we may henceforth speak of the *metric space* (R, d).

Now this metric space is *not necessarily complete*; i.e. we cannot say that for every Cauchy sequence $\{j_n\}$ in (R, d) (so that $d(j_n, j_m) < \epsilon$, $n, m \geqslant N_0$) there exists a limit point $z \in (R, d)$ such that $d(j_n, z) \to 0$ as $n \to \infty$. In other words (R, d) does not necessarily contain all its *limit points*. The following lemma will help us understand the process of making (R, d) complete, by transferring the problem to the real line.

LEMMA 5.10. A sequence $\{j_n\}$ is Cauchy in the metric space (R, d) if and only if the sequence of real numbers $K(i, j_n)$ is Cauchy (in respect to the usual metric on the real line) for each $i \in R$.

Proof. From the definition of $d(\,.\,,.\,)$ if $\{j_n\}$ is Cauchy in (R, d), since $\bar{u}_{1i}\, w_i > 0$, $\{K(i, j_n)\}$ is Cauchy on the real line for each i.

Conversely, if $\{K(i, j_n)\}$ is Cauchy for each i, let $\epsilon > 0$ and choose a finite subset of indices $E \subset R$ such that

$$\sum_{i \in R-E} w_i < \frac{\epsilon}{4}.$$

Choose M sufficiently large so that

$$\bar{u}_{1i}|K(i, j_m) - K(i, j_n)| < \frac{\epsilon}{2 \cdot \sum_{i \in R} w_i}$$

for each $i \in E$ and $n, m \geqslant M$.

Then writing symbolically

$$d(j_m, j_n) = \sum_{i \in E} + \sum_{i \in R-E}$$

in the definition of $d(\,.\,,.\,)$ we have

$$d(j_m, j_n) \leqslant \frac{\epsilon \sum_{i \in E} w_i}{2 \sum_{i \in R} w_i} + \frac{\epsilon}{2}$$

$$\leqslant \epsilon$$

for $n, m \geqslant M$. #

Suppose now that $\{j_n\}$ is any Cauchy sequence in (R, d). Then, by the result just proved, $\{K(i, j_n)\}$ is Cauchy for each fixed $i \in R$. Thus for each $i \in R$

$$\lim_{n \to \infty} K(i, j_n) = K(i, x)$$

where $K(i, x)$ is notation we adopt for the limit, which is of course some real number; of course $x \equiv x(\{j_n\})$ is not an entity we can 'picture' in general, but we only need to work with it insofar as we need only to know $K(i, x)$, $i \in R$. Define now for any $j \in R$

$$d(j, x) = \sum_{i \in R} |K(i, j_n) - K(i, x)|\bar{u}_{1i}\, w_i.$$

Then if $\{j_n\}$ is the Cauchy sequence corresponding to x

$$d(j_n, x) = \sum_{i \in R} |K(i, j_n) - K(i, x)|\bar{u}_{1i}\, w_i$$

$$\to 0$$

as $n \to \infty$, by dominated convergence. Thus if we *extend* the metric $d(\,.\,,.\,)$ from R to operate *also* on any *new* points x added in this way, in the obvious manner, and put $x = y$, if and only if $K(i, x) = K(i, y)$ for all $i \in R$, then we shall have an *extended,* and now *complete,* metric space (R^*, d). The set R is thus *dense* in (R^*, d), which is therefore *separable.* In fact more is now true:

LEMMA 5.11. The metric space (R^*, d) is compact.

Proof. [Compactness in metric space is equivalent to the Weierstrass property: that every infinite sequence has a limit point in the metric space.]

Let $\{k_n\}$ be any sequence in (R^*, d). We know that for each $i \in R$, and for any $k \in R$ and so (by obvious extension) for any $k \in R^*$,

$$0 \leqslant K(i, k) \leqslant 1/\bar{u}_{1i}.$$

Thus for any fixed $i \in R$, $K(i, k_n)$ is a bounded sequence of real numbers; it then follows from the Bolzano–Weierstrass Theorem and the Cantor diagonal refinement procedure that there exists a subsequence $\{k_{n_j}\}, j = 1, 2, \ldots$ of $\{k_n\}$ such that

$$\lim_{j\to\infty} K(i, k_{n_j}) \text{ exists for } \textit{every } i \in R.$$

Thus the sequence $K(i, k_{n_j})$ is Cauchy on the real line for every $i \in R$; and repeating now the argument in the second part of the proof of Lemma 5.10 ($k_{n_j} \in R^*$ now, and not necessarily to R) it follows that $\{k_{n_j}\}$ is Cauchy in (R^*, d) and, this last being complete, there exists a limit point ξ in (R^*, d) as required. #

DEFINITION 5.7. The set $R^* - R$ is called the Martin exit boundary of R, induced by the matrix P, *relative to index* 1.†

1. Now let $\mathcal{R}$ be the σ-field of Borel sets of (R^*, d), i.e. the minimum σ-field containing the open sets of (R^*, d). [It is obvious that each $i \in R$ is itself a closed set. Thus R itself is closed, and so $R^* - R$ is a Borel set.]

2. Further, consider $K(i, x)$ for fixed $i \in R$, as a function of $x \in (R^*, d)$. Then for $x_1, x_2 \in (R^*, d)$

$$d(x_1, x_2) < \delta \Rightarrow |K(i, x_1) - K(i, x_2)| < \epsilon$$

from the definition of the metric, where $\epsilon > 0$ is arbitrarily, and δ appropriately, chosen. Thus $K(i, x)$ is continuous on (R^*, d); and since (R^*, d) is *compact*, $K(i, x)$ is *uniformly continuous* on (R^*, d) for i fixed.

3. Let $\{P_n(\,.\,)\}$ be any sequence of probability measures‡ on the Borel sets $\mathcal{R}$ (i.e. measures such that $P_n(R^*) = 1$). Then since (R^*, d) is compact, there

† By suitable rearrangement of rows and columns of P, the development of the theory entails no essential loss in generality in working in relation to index 1.

‡ The assumption of probability measures is not strictly necessary to the sequel, but is convenient; especially for the reader acquainted with some probabilistic measure theory.

exists a subsequence $\{n_i\}$, $i \geq 1$ of the positive integers such that the subsequence $\{P_{n_i}(\,.\,)\}$ converges weakly to a limit probability measure $\{P(\,.\,)\}$. [This is the generalized analogue of the Helly selection principle for probability distributions on a closed line segment $[a, b]$. We shall use generalizations of another Helly theorem to compact metric space below. For proofs in the metric space setting, see Parthasarathy (1967), pp. 39–52; the theorem just used is on p. 45.]

THEOREM 5.8. Any superregular vector $\boldsymbol{u} = \{u(i)\}$ has representation

$$u(i) = \int_{R^*} K(i, x)\mu(\mathrm{d}x), \, i \in R$$

where $\mu(\,.\,)$ is some finite measure on $\mathcal{R}$ independent of i.

Proof. The case $\boldsymbol{u} = \boldsymbol{0}$ is trivial; assume $\boldsymbol{u} \neq \boldsymbol{0}$.

We shall first prove the proposition for $\boldsymbol{u} = \boldsymbol{h}$ where $\boldsymbol{h}$ is a finite non-zero potential. There exists a non-negative vector $\boldsymbol{k} = \{k(i)\}$ such that

$$\boldsymbol{h} = G\boldsymbol{k},$$

i.e.

$$h(i) = \sum_{j \in R} g_{ij}k(j)$$

$$= \sum_{j \in R} \frac{g_{ij}}{g_{1j}} g_{1j}k(j) = \sum_{j \in R} K(i, j)g_{1j}k(j).$$

Hence if we define a measure on $\mathcal{R}$ by

$$\mu(\{j\}) = g_{1j}k_j, \, j \in R, \qquad \mu(R^* - R) = 0,$$

then we have the required representation for $\boldsymbol{h}$, since

$$\mu(R^*) = \sum_{j \in R} g_{1j}k(j) = h(1);$$

$h(1) > 0$ since for all $j \in R$, $g_{1j} > 0$ by the Basic Assumption 2; and $k(j) > 0$ for at least one $j \in R$, as $\boldsymbol{h} \neq \boldsymbol{0}$ by assumption.

Now for a superregular vector $\boldsymbol{u} \neq \boldsymbol{0}$, let $\{\boldsymbol{h}_n\}$ be a non-decreasing sequence of potentials converging to $\boldsymbol{u}$ elementwise (by Theorem 5.7) and let the corresponding measures as deduced above be $\{\mu_n(\,.\,)\}$, where

$$\mu_n(R^*) = h_n(1) > 0.$$

We know $h_n(1) \uparrow u(1)$ as $n \to \infty$, so that $u(1) > 0$. Thus

$$h_n(i) = \int_{R^*} K(i, x)\mu_n(\mathrm{d}x)$$

where the sequence $\{\mu_n(\,.\,)\}$ does not depend on i.

Therefore
$$u(i) = \lim_{n\to\infty} h_n(i) = \lim_{n\to\infty} \int_{R^*} K(i, x)\mu_n(dx)$$
$$= \lim_{n\to\infty} h_n(1) \int_{R^*} K(i, x) \frac{\mu_n(dx)}{h_n(1)}$$

where $P_n(\,.\,) = \mu_n(\,.\,)/h_n(1)$ is a probability measure on $\mathcal{R}$. Hence by the generalized Helly selection principle mentioned in 3 above, and by the generalized Helly–Bray Lemma, since $K(i, x)$ is bounded and continuous in $x \in R^*$, for fixed $i \in R$

$$u(i) = u(1) \int_{R^*} K(i, x)P(dx)$$

where $P(\,.\,)$ is the limit probability measure on $\mathcal{R}$ of an appropriate subsequence $\{P_{n_i}(\,.\,)\}$.

The theorem now follows by putting

$$\mu(\,.\,) = u(1)P(\,.\,). \quad \#$$

5.6. Example†

We consider the infinite stochastic matrix P, defined on the index set $R = \{1, 2, \ldots\}$ by

$$p_{i,i+1} = p_i \; (>0)$$
$$p_{i,1} = 1 - p_i = q_i \; (>0), \; i \in R.$$

Thus we may represent P in the form

$$P = \begin{bmatrix} q_1 & p_1 & 0 & 0 & \ldots \\ q_2 & 0 & p_2 & 0 & \\ q_3 & 0 & 0 & p_3 & \\ \vdots & & & & \ddots \end{bmatrix}.$$

The index set R clearly forms a single essential aperiodic class, and hence P is primitive.

For $k \geqslant 1$

$$l_{11}^{(k)} = f_{11}^{(k)} = p_0 p_1 \ldots p_{k-1} q_k, \quad (p_0 = 1)$$
$$= (p_0 p_1 \ldots p_{k-1}) - (p_0 p_1 \ldots p_k)$$

since $q_k = 1 - p_k$. Hence

$$L_{11}(1-) \equiv F_{11}(1-) \equiv \sum_{k=1}^{\infty} f_{11}^{(k)} = p_0 - \lim_{k\to\infty} (p_0 p_1 \ldots p_k)$$
$$= p_0 - \lim_{k\to\infty} \alpha_k$$

† See Example (4) of §4.1 for another view of this example.

where $\alpha_k = p_0 p_1 \ldots p_k$ has a limit $\alpha_\infty \geqslant 0$ as $k \to \infty$ since it is positive and decreasing with k. Hence

$$L_{11}(1-) = F_{11}(1-) = 1 - \alpha_\infty .$$

Thus index 1 (and thus every index) is transient if and only if $\alpha_\infty > 0$ (and is recurrent if $\alpha_\infty = 0$).

In the case $\alpha_\infty = 0$:

$$\begin{aligned}\mu &= \sum_{k=1}^{\infty} k f_{11}^{(k)} = \sum_{k=1}^{\infty} k(\alpha_{k-1} - \alpha_k) \\ &= \sum_{k=1}^{\infty} \sum_{j=1}^{k} (\alpha_{k-1} - \alpha_k) \\ &= \sum_{j=1}^{\infty} \sum_{k=j}^{\infty} (\alpha_{k-1} - \alpha_k) = \sum_{j=1}^{\infty} \alpha_{j-1} .\end{aligned}$$

Hence

$$\mu = \sum_{j=0}^{\infty} \alpha_j ,$$

and we obtain positive recurrence of the matrix P if $\sum_{j=0}^{\infty} \alpha_j < \infty$ and null-recurrence otherwise (when $\alpha_\infty = 0$).

We shall next evaluate in terms of the α_is the quantities $F_{ij}(1-)$, $i \neq j$; this is really of interest only when $\alpha_\infty > 0$, i.e. in the case of transience of the matrix P, for otherwise† $F_{ij}(1-) = 1$, all $i, j \in R$, but we shall not assume that $\alpha_\infty > 0$ necessarily. [The case $F_{ii}(1-)$, $i \in R$ is – at least in the transient case – more difficult to evaluate, but, as we shall see, we shall not need if for consideration of the Martin exit boundary.]

First we note that

$$F_{ij}(1-) = 1 \text{ if } 1 \leqslant i < j$$

for this quantity is readily seen to be the 'absorption probability' from index i of the set $1 \leqslant i < j$ [to which there corresponds a strictly substochastic primitive matrix ${}_{(j-1)}P$ (the $(j-1) \times (j-1)$ northwest corner truncation of P)] into the single 'absorbing index' of remaining indices $\{j, j+1, \ldots\}$. (See Exercise 4.12.)

Secondly

$$F_{ij}(1-) = F_{i1}(1-)F_{1j}(1-), \text{ if } i > j \geqslant 1$$

which follows from the definition of $F_{ij}(1-)$ in this special situation, i.e. in view of the previous result

$$F_{ij}(1-) = F_{i1}(1-), \text{ if } i > j \geqslant 1.$$

† See Exercise 5.4 for the general proposition.

Thus we need only to find $F_{i1}(1-)$ for $i>1$. Now

$$f_{i1}^{(k)} = q_i,\ k = 1$$
$$= p_i p_{i+1} \ldots p_{i+k-2} q_{i+k-1},\ k \geqslant 2$$

from the special structure of P.

Hence

$$F_{i1}(1-) = q_i + \sum_{k=2}^{\infty} [p_i p_{i+1} \ldots p_{i+k-2} - p_i p_{i+1} \ldots p_{i+k-1}]$$
$$= q_i + p_i - \lim_{k\to\infty} p_i p_{i+1} \ldots p_{i+k}$$
$$= q_i + p_i - \lim_{k\to\infty} \frac{p_0 p_1 \ldots p_{i+k}}{p_0 p_1 \ldots p_{i-1}}$$
$$= 1 - \frac{\alpha_\infty}{\alpha_{i-1}}.$$

Thus, finally

$$F_{ij}(1-) = \begin{cases} 1 & \text{if } i<j, \\ 1 - \dfrac{\alpha_\infty}{\alpha_{i-1}} & \text{if } i>j\geqslant 1. \end{cases}$$

Let us now pass on to the Martin boundary theory, assuming transience henceforth i.e. $\alpha_\infty > 0$.

Then

$$\boxed{\begin{aligned} \bar{u}_{ij} &= 1 && \text{if } 1 \leqslant i \leqslant j \\ &= 1 - \frac{\alpha_\infty}{\alpha_{i-1}} && \text{if } i>j, \end{aligned}}$$

so that

$$K(i,j) = \frac{\bar{u}_{ij}}{\bar{u}_{1j}} = \bar{u}_{ij}.$$

Hence (relative to index 1 as usual) we have the metric on $(R \times R)$

$$d(\nu_1, \nu_2) = \sum_{i\in R} w_i \bar{u}_{1i} |K(i, \nu_1) - K(i, \nu_2)|$$
$$= \sum_{i\in R} w_i |K(i, \nu_1) - K(i, \nu_2)|.$$

Assume, for further working, that we are dealing with $\nu_1 \neq \nu_2$; so without loss of generality, let us assume $\nu_1 < \nu_2$; then

$$d(\nu_1,\nu_2) = \sum_{i=1}^{\nu_1} w_i(1-1) + \sum_{i=\nu_1+1}^{\nu_2} w_i \left|1 - \frac{\alpha_\infty}{\alpha_{i-1}} - 1\right| + \sum_{i=\nu_2+1}^{\infty} w_i \left|1 - \frac{\alpha_\infty}{\alpha_{i-1}} - \left(1 - \frac{\alpha_\infty}{\alpha_{i-1}}\right)\right|,$$

i.e. for $\nu_1 < \nu_2$

$$d(\nu_1, \nu_2) = \sum_{\nu_1 < i \leqslant \nu_2} w_i \frac{\alpha_\infty}{\alpha_{i-1}}$$

where, of course, $0 < \alpha_\infty/\alpha_{i-1} < 1$.

Thus the metric space (R, d) is *isometric* to the metric space which is that subset of the real line (with the ordinary modulus-difference metric) consisting of points

$$\nu_1, \nu_1 + \nu_2, \nu_1 + \nu_2 + \nu_3, \ldots$$

where
$$\nu_i = w_i\, \alpha_\infty/\alpha_{i-1}, i \in R.$$

Now every *non-terminating* subsequence of this new metric space converges to the same point, viz.

$$\sum_{i=1}^{\infty} \nu_i \equiv \sum_{i=1}^{\infty} w_i\, \alpha_\infty/\alpha_{i-1} < \infty.$$

Hence the Cauchy completion of (R, d), viz. the Martin exit boundary of (R, d) relative to index 1 induced by the matrix P, consists of a *single point,* which we may call, therefore, 'infinity'.

Bibliography and discussion

Our discussion of countable stochastic matrices is rather selective; excellent treatments now exist in textbook and monograph form, and the reader desiring a sustained and more extensive development of most of the material presented in this chapter, as well as extensive bibliography, should consult the books of Feller (1968), Chung (1967) and Kemeny, Snell & Knapp (1966), expecially Chung's.

As mentioned earlier in the book, study of countable stochastic matrices was initiated by Kolmogorov (1936), by whom the notions of essential and inessential indices, which we use throughout the book, were introduced; and followed closely by many contributions of Doeblin, although we have not touched to any significant extent on this later author's work in this context.

The standard approach to the basic theory is in terms of the quantities $\{f_{ij}^{(k)}\}$. We have worked in terms of the quantities $\{\ell_{ij}^{(k)}\}$ partly so as not to duplicate this approach yet again, although for stochastic P it is in several ways not quite so natural as the other. The fundamental reason, however, for using the $\{\ell_{ij}^{(k)}\}$ is that even in the stochastic context it is the natural tool as regards the development of the probabilistically important theory of sub-invariant and invariant measures, and makes a reasonably unified treatment possible; whereas, e.g. in Feller (1968), a change of approach is necessary when passing on to the topic of invariant measures. Finally, it is necessary to mention that in §§5.1, 5.3 and 5.4, the present author was substantially influenced in approach by the papers of Vere–Jones (1962, 1967), which are not restricted to the stochastic framework, and play a substantial role in the development of the next chapter also.

Theorem 5.1′ is due to Erdös, Feller & Pollard (1949). The part of Theorem 5.4 referring to recurrent P is due to Derman (1954). A necessary and sufficient condition for the existence of an invariant measure for a transient P is due to Harris (1957) and Veech (1963). Theorem 5.5 is due to Kolmogorov (1936) and Feller (1968; 1st ed. 1950). For a probabilistic rather than analytical proof of Theorem 5.5 the reader should consult Orey (1962).

A consideration of Perron–Frobenius-type spectral properties of denumerable stochastic irreducible P as operators on various sequence spaces (i.e. from a functional-analytic point of view) has been give by Nelson (1958), Kendall (1959), Sidák (1962, 1963, 1964*a*) (see also Holmes (1966)), and Moy (1965), and these results provide an interesting comparison with those obtained in the present chapter. The reader interested in this topic should also consult the references in the notes at the conclusion of the next chapter, which refer to a more general (specifically, non-stochastic) situation.

Much of §5.5 is adapted from the papers of Doob (1959) and Moy (1967*a*). Extensive discussions of the potential theory in the same framework as in §5.5 may be found in the book of Kemeny, Snell & Knapp (1966) and the earlier exposition of Neveu (1964).

Exercises

5.1. Let i be an index with period $d > 1$ corresponding to the stochastic matrix $P = \{p_{ij}\}$. Show that in relation to the matrix

$$P^d = \{p_{ij}^{(d)}\},$$

i is aperiodic, and show also that for $0 \leq s < 1$,

$$L_{ii}(s) = L_{ii}^{(d)}(s^{1/d})$$

by first showing

$$P_{ii}(s) = P_{ii}^{(d)}(s^{1/d})$$

where the superscript (d) pertains to the matrix P^d.

Use this to show that if i is positive or null recurrent, or transient in relation to P, it is also this, respectively, in relation to P^d; and that in the recurrent case, as $k \to \infty$

$$p_{ii}^{(kd)} \to d/\mu_i.$$

5.2. Use the final result of Exercise 5.1 to conclude that for an irreducible stochastic matrix P (not necessarily aperiodic) for all $i, j = 1, 2, \ldots$

$$p_{ij}^{(k)} \to 0$$

if the matrix P is transient, or null-recurrent; and that this occurs for *no* pair (i, j) if P is positive recurrent.

5.3. Show, if i is an index corresponding to a *finite* stochastic matrix P, that i is transient if and only if i is inessential; and is positive recurrent otherwise.

5.4. Develop in full the 'dual' approach to the theory as indicated in §5.4. Prove in particular the analogues of Theorems 5.3 and 5.4. Prove also Theorem 5.5 from this new viewpoint, noting in the course of this that for a recurrent stochastic P, $F_{ij}(1-) = 1$ for all $i, j = 1, 2, \ldots$

5.5. Show that a stochastic irreducible matrix P is positive-recurrent if and only if there exists a *subinvariant* measure x' for P such that $x'\mathbf{1} < \infty$. Repeat for an invariant measure.

5.6. An infinite stochastic irreducible $P = \{p_{ij}\}$ satisfies

$$\sum_{j=1}^{N} p_{ij} \geqslant \delta > 0$$

uniformly for all i in the index set of P, where N is some fixed positive integer. Show that P is positive-recurrent.

Hint: Consider $$\sum_{j=1}^{N} p_{ij}^{(k)}.$$

5.7. For the Example to which §5.6 is devoted, attempt to solve the invariant equations $x'P = x'$, and hence deduce a necessary and sufficient condition for positive-recurrence of this particular P, using the result of Exercise 5.5. Does this concur with the results of §5.6 on classification?

Find μ_i for each $i = 1, 2, \ldots$ in the positive-recurrent case.

5.8. Let $P = \{p_{ij}\}$ and $P_n = \{p_{ij}(n)\}$, $n = 1, 2, \ldots$ be transition probability matrices of irreducible recurrent Markov chains each defined on the states $\{1, 2, \ldots\}$, and let $\{\nu_i\}$ and $\{\nu_i^{(n)}\}$, $n = 1, 2, \ldots$ be the corresponding invariant measures, normalized so that $\nu_1 = \nu_1^{(n)} = 1$ for all n. Assume that $p_{ij}(n) \to p_{ij}$ for all i, j as $n \to \infty$.

Show that $\{\nu_i^*\}$, defined by $\nu_i^* = \liminf_{n\to\infty} \nu_i^{(n)}$, $i = 1, 2, \ldots$, is a subinvariant measure corresponding to P; and hence show that

$$\nu_i^* = \lim_{n\to\infty} \nu_i^{(n)}, \; i = 1, 2, \ldots$$

Finally, deduce that $\nu_i = \nu_i^*$, $i = 1, 2, \ldots$

5.9. If P is an infinite recurrent doubly stochastic matrix, show that it must be null-recurrent.

5.10. Show that the structure of the Martin exit boundary $B = R^* - R$ does not depend on the choice of weights $\{w_i\}$ in the definition of the metric $d(\nu_1, \nu_2)$ on $(R \times R)$, so long as $w_i > 0$ all i, $\sum_{i\in R} w_i < \infty$. (Use Lemma 5.10.)

5.11. Let $P = \{p_{ij}\}$ be an irreducible substochastic matrix on $R \times R$ with one index transient. Show that all $i \in R$ are also transient, and that $x' = \{x_i\}$ defined by $x_i = g_{1i}/g_{11}$ is a strictly positive superregular *row* vector for P.

Further show that the matrix $\hat{P} = \{\hat{p}_{ij}\}$, where

$$\hat{p}_{ij} = x_j p_{ji}/x_i, \; i, j \in R$$

is irreducible, substochastic, and all its indices are transient.

Hint: Show $$\hat{g}_{ij} = x_j g_{ji}/x_i.$$

The Martin *entrance* boundary (relative to index 1) for matrix P is the Martin exit boundary for the matrix $\hat{P}$.

Show that† for $i, j \in R$

$$\hat{K}(i, j) = K(j, i)/K(j, 1).$$

5.12. Using the results of Exercise 5.11 in relation to P as given in § 5.6, with $\alpha_\infty > 0$, show that

$$\hat{K}(i, j) = \begin{cases} 1 & , \; j = 1, i \geq 1 \\ \left[1 - \dfrac{\alpha_\infty}{\alpha_{j-1}}\right]^{-1} & , \; 1 < j \leq i \\ 1 & , \; j > i \geq 1 \end{cases}$$

and hence that the space $(\hat{R}, \hat{d})$ is *itself complete*, so that the Martin *entrance* boundary of P, corresponding to index 1, is empty.

5.13. The *substochastic* irreducible matrix $P = \{p_{ij}\}$ on $R = \{1, 2, \ldots\}$ described by

$$p_{j,j+1} = p_j > 0, \; p_{j,j-1} = q_j > 0, \; p_{j,j} = 1 - p_j - q_j \geq 0$$

for $j \geq 1$, describes the 'transitions' between the inessential indices (those of R) of a (space inhomogeneous) random walk on the non-negative integers with 'absorbing barrier' at the 'origin', 0. For this matrix, it can be shown‡ that $G = \{g_{ij}\}$ is given by

$$g_{ij} = \begin{cases} \bar{q}_i \dfrac{\sum_{s=0}^{j-1} \rho_s}{q_j \rho_{j-1}}, & 1 \leq j \leq i \\[2ex] \bar{q}_j \dfrac{\sum_{s=0}^{i-1} \rho_s}{q_j \rho_{j-1}}, & 1 \leq i \leq j \end{cases}$$

where

$$\bar{q}_i = \frac{\sum_{s=i}^{\infty} \rho_s}{\sum_{s=0}^{\infty} \rho_s} \quad \rho_0 = 1, \; \rho_i = \frac{q_1 q_2 \ldots q_i}{p_1 p_2 \ldots p_i}$$

† Capped symbols refer to the situation in relation to matrix $\hat{P}$.

‡ Seneta (1967*b*).

and $\bar{q}_i$ (the 'absorption probability' into the origin from $i \geqslant 1$) is to be understood as unity if $\Sigma \rho_s$ diverges.

Calculate $K(i, j)$, $i, j \in R$, and hence deduce the structure of the Martin exit boundary.

Repeat with the Martin entrance boundary, using the comments of Exercise 5.11, and $\hat{K}(i, j)$.

6 Countable non-negative matrices

Countable stochastic matrices P, studied in the previous chapter, have two (related) essential properties not generally possessed by countable non-negative matrices $T = \{t_{ij}\}, i, j = 1, 2, \ldots$ In the first instance the powers P^k, $k \geq 1$, are all well defined (using the obvious extension of matrix multiplication); secondly the matrix P (and its powers) have row sums unity.

It is inconvenient in the general case to deal with (as usual elementwise finite) matrices T whose powers may have infinite entries. For the sequel we therefore make the

Basic Assumption 1: $T^k = \{t_{ij}^{(k)}\}$, $k \geq 1$, are all elementwise finite.

The second role played by the 'row-sums unity' assumption involved in the stochasticity of P, is that this last may be expressed in the form $P\mathbf{1} = \mathbf{1}$, whence we may expect that even in the case where P is actually infinite, our study of Perron–Frobenius-type structure of P may still be centred about the 'eigenvalue' unity, and the 'right eigenvector' playing a role similar to the finite case will be $\mathbf{1}$. The reader examining the details of the last chapter will see that this is in fact the approach which was adopted.

In the case of general non-negative T the asymmetric stochasticity assumption is absent, and we may, in the first instance, expect to need to resolve the problem (restricting ourselves to the irreducible case) of the natural analogue of the Perron–Frobenius eigenvalue. Indeed the resolution of this problem, and associated problems concerning 'eigenvectors', makes this theory more fundamental even for transient stochastic P (as will be seen), then that of the last chapter.

For convenience we shall confine ourselves to those T which satisfy the

Basic Assumption 2: T is irreducible.

Recall that in the countable context irreducibility is defined precisely as in the finite case, viz. for each $i, j = 1, 2, \ldots$ there exists $k \equiv k(i, j)$ such that $t_{i,j}^{(k)} > 0$. It also follows as in Chapter 1, that all indices of T have a common finite period $d \geq 1$ (for further detail of this kind see §5.1).

6.1 The convergence parameter R, and the R-classification of T.

THEOREM 6.1. The power series

$$T_{ij}(z) = \sum_{k=0}^{\infty} t_{i,j}^{(k)} z^k, \; i, j = 1, 2, \ldots$$

all have common convergence radius R, $0 \leqslant R < \infty$, for each pair i, j.

Proof. Denote the convergence radius of $\sum_{k=0}^{\infty} t_{ij}^{(k)} z^k$ by R_{ij}. On account of the non-negativity of the coefficients in the power series,

$$R_{ij} = \sup_{s \geqslant 0} (s : \sum_k t_{ij}^{(k)} s^k < \infty).$$

We now consider the obvious inequalities, frequently used hitherto,

$$t_{ij}^{(\nu+k)} \geqslant t_{ij}^{(\nu)} t_{jj}^{(k)}$$

$$t_{jj}^{(\nu+k)} \geqslant t_{ji}^{(\nu)} t_{ij}^{(k)}$$

$$t_{ii}^{(\nu+k)} \geqslant t_{ij}^{(k)} t_{ji}^{(\nu)}$$

$$t_{ij}^{(\nu+k)} \geqslant t_{ii}^{(k)} t_{ij}^{(\nu)}.$$

The first inequality clearly implies that

$$R_{ij} \leqslant R_{jj}$$

(consider ν constant and such that $t_{ij}^{(\nu)} > 0$, using irreducibility; and form the relevant power series), while the second analogously implies that

$$R_{jj} \leqslant R_{ij}.$$

Thus we have that $R_{ij} = R_{jj}$ for all $i, j = 1, 2, \ldots$ The third and fourth inequalities imply, respectively, that

$$R_{ii} \leqslant R_{ij}, \; R_{ij} \leqslant R_{ii}$$

so that $R_{ij} = R_{ii}$ for all $i, j = 1, 2, \ldots$

Thus we may write R for the common value of the R_{ij}.

It remains to show only that $R < \infty$. This can be seen via Lemma A.4 of Appendix A, which implies, in view of

$$t_{ii}^{(k+r)d} \geqslant t_{ii}^{(kd)} t_{ii}^{(rd)}, \; k, r \geqslant 0$$

where d is the period of t, that

$$\lim_{n\to\infty} \{t_{ii}^{(nd)}\}^{1/n}$$

exists and is positive and so

$$\lim_{n\to\infty} \{t_{ii}^{(nd)}\}^{1/nd} \equiv \limsup_{k\to\infty} \{t_{ii}^{(k)}\}^{1/k}$$

exists and is positive, and by a well-known theorem† the limit is $1/R_{ii} \equiv 1/R$.#

† e.g. Titchmarsh (1939, §7.1)

From Lemma A.4 of Appendix A we obtain the additional information:

COROLLARY. $t_{ii}^{(nd)} \leqslant R^{-nd}$

i.e.

$$R^k \, t_{ii}^{(k)} \leqslant 1$$

for all $i = 1, 2, \ldots, k \geqslant 0$; and

$$\lim_{n\to\infty} \{t_{ii}^{(nd)}\}^{1/nd} = R^{-1}.$$

N.B. It is possible to avoid the use of Lemma A.4 altogether here,† through an earlier introduction of the quantities $F_{ij}(z)$, $L_{ij}(z)$ which we carry out below; its use has been included for unity of exposition, and for the natural role which it plays in this theory.

DEFINITION 6.1. The common convergence radius, R, $0 \leqslant R < \infty$, of the power series

$$\sum_{k=0}^{\infty} t_{ij}^{(k)} z^k$$

is called the *convergence parameter* of the matrix T.

Basic Assumption 3: We deal only with matrices T for which $R > 0$.

Note that for a countable substochastic (including stochastic) matrix $T = P, R \geqslant 1$, clearly. Further, it is clear, if T is finite that $1/R = r$, the Perron–Frobenius eigenvalue of T (in the aperiodic case this is obvious from Theorem 1.2 of Chapter 1, and may be seen to be so in the period d case by considering a primitive class of T^d). For this reason the quantity $1/R$ is sometimes used in the general theory, and bears the name of *convergence norm* of T.

To proceed further, we define quantities $f_{ij}^{(k)}$, $\ell_{ij}^{(k)}$ as in the last chapter; write $f_{ij}^{(0)} = \ell_{ij}^{(0)} = 0$, $f_{ij}^{(1)} = \ell_{ij}^{(1)} = t_{ij}$ and thereafter write inductively

$$f_{ij}^{(k+1)} = \sum_{\substack{r \\ r\neq j}} t_{ir} f_{rj}^{(k)}, \quad \ell_{ij}^{(k+1)} = \sum_{\substack{r \\ r\neq i}} \ell_{ir}^{(k)} t_{rj} \qquad i, j = 1, 2, \ldots$$

Clearly $f_{ij}^{(k)}, \ell_{ij}^{(k)} \leqslant t_{ij}^{(k)}$, and so

$$F_{ij}(z) = \sum_{k=0}^{\infty} f_{ij}^{(k)} z^k, \quad L_{ij}(z) = \sum_{k=0}^{\infty} \ell_{ij}^{(k)} z^k$$

are convergent for $|z| < R$, and as in the previous chapter

$$T_{jj}(z)\, F_{ij}(z) = T_{ii}(z)\, L_{ij}(z) = T_{ij}(z), \; i \neq j;$$

$$L_{ii}(z) = F_{ii}(z); \; T_{ii}(z) = (1 - L_{ii}(z))^{-1}$$

for $|z| < R$; the very last equation follows from

$$T_{ii}(z) - 1 = T_{ii}(z)\, L_{ii}(z)$$

† See Exercise 6.1.

for $|z| < R$, through the fact that for $|z| < R$

$$|L_{ii}(z)| \leq L_{ii}(|z|)$$

and for $0 \leq s < R$

$$L_{ii}(s) = 1 - [T_{ii}(s)]^{-1} < 1$$

as before. We thus have, letting $s \to R-$, that

$$L_{ii}(R-) \equiv F_{ii}(R-) \leq 1.$$

Finally let us put

$$\mu_i(R) = R\, L'_{ii}(R-) = \sum_k k\, \ell_{ii}^{(k)} R^k.$$

DEFINITION 6.2. An index i is called R- *recurrent* if $L_{ii}(R-) = 1$ and R- *transient* if $L_{ii}(R-) < 1$.

An R- *recurrent* index i is said to be R- *positive or* R- *null* depending as $\mu_i(R) < \infty$ or $\mu_i(R) = \infty$ respectively.

Clearly, an index i is R- transient if and only if $T_{ii}(R-) < \infty$; in this case $t_{ii}^{(k)} R^k \to 0$ as $k \to \infty$. To go further, we can adapt the results of Chapter 5 for our convenience in the following way: proceeding as in Lemma 5.4, we obtain for $|z| < R$ that

$$L_{ij}(z) = z \sum_r L_{ir}(z)\, t_{rj} + z\, t_{ij}\, (1 - L_{ii}(z))$$

which yields eventually that $L_{ij}(R-) < \infty$ for all i, j ($i \leftrightarrow j$ because of irreducibility) and as in Lemma 5.5 we obtain that there is always a row vector $\boldsymbol{x}'$, $\boldsymbol{x}' \geq \boldsymbol{0}'$, $\neq \boldsymbol{0}'$ satisfying

$$R\, \boldsymbol{x}'\, T \leq \boldsymbol{x}' \tag{6.1}$$

one such being given by the vector $\{L_{ij}(R-)\}$, $j = 1, 2, \ldots$ [for arbitrary fixed i].

DEFINITION 6.3. Any $\boldsymbol{x}' \geq \boldsymbol{0}'$, $\neq \boldsymbol{0}'$ satisfying

$$R\, \boldsymbol{x}'\, T \leq \boldsymbol{x}'$$

is called an R- *subinvariant* measure. If in fact $R\, \boldsymbol{x}'\, T = \boldsymbol{x}'$, $\boldsymbol{x}'$ is called an R- *invariant* measure.

As in Lemma 5.6 any R- subinvariant measure has all its entries positive.

Now for a given R- subinvariant measure $\boldsymbol{x}' = \{x_i\}$ define an, in general, substochastic matrix $P = \{p_{ij}\}$ by

$$p_{ij} = R\, x_j\, t_{ji}/x_i, \; i, j = 1, 2, \ldots$$

so that $P^k = \{p_{ij}^{(k)}\}$ has its elements given by

$$p_{ij}^{(k)} = R^k\, x_j\, t_{ji}^{(k)}/x_i.$$

It is readily checked P is irreducible. Let us then consider two cases.

1. P is strictly substochastic, i.e. at least one row sum is less than unity. Then as at the outset of §5.5, each of the indices $i = 1, 2, \ldots$ may be considered inessential in an enlarged matrix $\bar{P}$, hence transient from Lemma 5.2, whence for each $i = 1, 2, \ldots$

$$\sum_k p_{ii}^{(k)} < \infty. \tag{6.2}$$

We shall in the rest of this section include this case with the case where P is *stochastic and transient,* in which case (6.2) holds also.

2. P is stochastic, in which case it may be transient, null-recurrent or positive-recurrent.

It can be seen that the transformation will enable us to fall back, to a large extent, on the results presented in Chapter 5 by considering the matrix P, and then translating the results for the matrix T.

Thus for example an index i of P is transient if and only if

$$\sum_{k=0}^{\infty} p_{ii}^{(k)} < \infty; \Leftrightarrow \sum_{k=0}^{\infty} t_{ii}^{(k)} R^k < \infty,$$

i.e. if and only if i in T is R- transient. Moreover since P (like T) is irreducible, all its indices are either transient or recurrent; *thus all indices of T must be R- transient, or all must be R- recurrent.* The last case can only occur if P is stochastic.

Further note that for $|z| < 1$

$$[1 - L_{ii}^{(P)}(z)]^{-1} = P_{ii}(z) = \sum_{k=0}^{\infty} p_{ii}^{(k)} z^k = \sum_{k=0}^{\infty} t_{ii}^{(k)} R^k z^k$$

$$= [1 - L_{ii}^{(T)}(Rz)]^{-1}$$

where the superscripts P and T indicate which matrix is being referred to. Hence for $0 \leq s < 1$

$$\frac{1 - L_{ii}^{(P)}(s)}{1 - s} = \frac{1 - L_{ii}^{(T)}(Rs)}{1 - s}$$

and letting $s \to 1-$ in the R- recurrent case of T (= the recurrent case of P)

$$L_{ii}^{(P)\prime}(1-) = R\, L_{ii}^{(T)\prime}(R-).$$

Thus i in P is positive recurrent if and only if i in T is R- positive, etc., and since all i in recurrent P are positive recurrent, or all are null-recurrent, for *R- recurrent T all indices are R- positive or R- null.* Thus it is now sensible to make the:

DEFINITION 6.4. T is called R- transient, R- positive or R- null if one of its indices is R- transient, R- positive or R- null respectively.

It is clear, further, that if T is R- transient or R- null, for any i, j, $R^k\, t_{ij}^{(k)} \to 0$ as $k \to \infty$.

If T is R- positive,

$$R^k\, t_{ij}^{(k)}$$

is clearly still bounded as $k \to \infty$ for any fixed pair of indices (i, j) since

$$x_i\, R^k\, t_{ij}^{(k)}/x_j = p_{ji}^{(k)} \leqslant 1.$$

On the other hand, no matter what the R- classification of T,

$$\beta^k\, t_{ij}^{(k)}$$

is unbounded as $k \to \infty$ for any pair (i, j), if $\beta > R$. For suppose not.

Then
$$\beta^k\, t_{ij}^{(k)} \leqslant K_{ij} = \text{const.}$$

for all k. Let z satisfy $|z| < \beta$. Then clearly

$$T_{ij}(z) = \sum_{k=0}^{\infty} z^k\, t_{ij}^{(k)}$$

converges for such z and hence converges for z satisfying $\beta > |z| > R$, which are outside the radius of convergence of $T_{ij}(z)$; which is a contradiction.

In conclusion to this section we note that the *common convergence radius* of all $T_{ij}(z)$, and the *common R- classification of all indices* in an irreducible T are further examples of so-called *solidarity* properties, enjoyed by all indices of an irreducible non-negative matrix (another example is the common period d of all indices).

6.2 R- subinvariance and invariance; R- positivity

As in Theorem 5.3 if $\boldsymbol{x}' = \{x_i\}$ is any R- subinvariant measure corresponding to our (irreducible, $R > 0$) T, then for fixed but arbitrary i, and all $j = 1, 2, \ldots$

$$x_j/x_i \geqslant \bar{x}_{ij}$$

where
$$\bar{x}_{ij} = (1 - \delta_{ij})\, L_{ij}\, (R\text{-}) + \delta_{ij}$$

and $\{\bar{x}_{ij}\}$, $j = 1, 2, \ldots$ is also an R- subinvariant measure [with ith element unity]. We then have in the same manner as before for Theorem 5.4:

THEOREM 6.2. For an R- recurrent matrix T an R- invariant measure always exists, and is a constant multiple of $\{\bar{x}_{ij}\}$, $j = 1, 2, \ldots$

An R- subinvariant measure which is not invariant exists if and only if T is R- transient; one such is then given by $\{\bar{x}_{ij}\}$, $j = 1, 2, \ldots$

N.B. The analogous discussion with the quantities $F_{ij}(R\text{-})$ would not be, in the present context, essentially different from the above, since there is no 'asymmetric' assumption such as stochasticity present, which tends to endow T and T' with somewhat divergent properties. (In any case it is easily shown that the common convergence radius R is the same for T and T', and the subinvariant *vector* properties of T may be evolved from the subinvariant *measure* properties of T'.)

We now trivially extend the scope of our investigations of subinvariance, by saying:

DEFINITION 6.5. $x' = \{x_i\}, x' \geqslant 0', \neq 0'$ is a β- *subinvariant* measure for $\beta > 0$, if

$$\beta x' T \leqslant x'$$

elementwise. The definition of β- invariance is analogous.

THEOREM 6.3. (*a*) If $x' = \{x_i\}$ is a β- subinvariant measure, then $x > 0$ and

$$x_j/x_i \geqslant \delta_{ij} + (1 - \delta_{ij}) L_{ij}(\beta);$$

(*b*) $L_{ii}(\beta) \leqslant 1$ for all i and $\beta \leqslant R$;

(*c*) for $\beta \leqslant R$, $L_{ij}(\beta) < \infty$ for all i, j and for fixed i $\{L_{ij}(\beta)\}$ constitutes a left β- subinvariant meausre, remaining subinvariant if $L_{ii}(\beta)$ is replaced by unity;

(*d*) no β-subinvariant measure can exist for $\beta > R$.

Proof. In view of the theory of this chapter and Chapter 5, the only proposition which is not clear from already established methods is (*d*), and we shall prove only this.

Suppose $x' \geqslant 0', \neq 0'$ is a β- subinvariant measure for $\beta > R$. Then since

$$\beta^k x' T^k \leqslant x'$$

we have for a $\bar{\beta}$ such that $R < \bar{\beta} < \beta$

$$x' \left(\sum_{k=0}^{\infty} (\bar{\beta} T)^k \right) \leqslant (1 - \bar{\beta}|\beta)^{-1} x'.$$

But, since $\bar{\beta} > R$ the convergence radius of $\sum_{k=0}^{\infty} t_{ij}^{(k)} z^k$, the left hand side cannot be elementwise finite, which is a contradiction. #

We now pass on to a deeper study of the interaction of β- subinvariance and invariance properties as they relate to the important case of R- positivity of T.

THEOREM 6.4. Suppose $x' = \{x_i\}$ is a β- invariant measure and $y = \{y_i\}$ β- invariant vector of T. Then T is R- positive if

$$y'x = \sum_i y_i x_i < \infty,$$

in which case $\beta = R$, x' is (a multiple of) the unique R- invariant measure of T and y is (a multiple of) the unique R- invariant vector of T.

Conversely, if T is R- positive, and x', y are respectively an invariant measure and vector, then $y'x < \infty$.

Proof. We have

$$\beta \mathbf{x}'T = \mathbf{x}', \beta T\mathbf{y} = \mathbf{y}$$

where $0 < \beta$, and $\beta \leqslant R$, the last following from Theorem 6.3 (*d*), [and $\mathbf{x} > \mathbf{0}, \mathbf{y} > \mathbf{0}$ from Theorem 6.3 (*a*)]. Form the stochastic matrix $P = \{p_{ij}\}$ where

$$p_{ij} = \beta x_j t_{ji}/x_i.$$

Suppose first $\sum_i x_i y_i < \infty$, and norm so that $\sum_i x_i y_i = 1$; and put $\boldsymbol{u} = \{v_i\}$, where $v_i = x_i y_i$. Then

$$\sum_i v_i p_{ij} = x_j \beta \sum_i t_{ji} y_i = x_j y_j = v_j$$

for each i. Thus

$$\boldsymbol{\nu}'P = \boldsymbol{\nu}'$$

and $\boldsymbol{\nu}'\mathbf{1} = 1$. Thus irreducible stochastic P has a stationary distribution, and by the Corollary to Theorem 5.5 this is possible only if P is positive-recurrent.

Hence

$$p_{ij}^{(k)} = \beta^k t_{ji}^{(k)} x_j/x_i \nrightarrow 0$$

as $k \to \infty$; since for $\beta < R$, $\beta^k t_{ji}^{(k)} \to 0$, from the definition of R, and as we know $\beta \leqslant R$, it must follow $\beta = R$, and that T is in fact R- positive, for otherwise $R^k t_{ji}^{(k)} \to 0$ as $k \to \infty$. The rest follows from Theorem 6.2.

Conversely, assume T is R- positive, and $\mathbf{x}'$ and $\mathbf{y}$ are respectively an invariant measure and vector. Form the stochastic matrix $P = \{p_{ij}\}$ where

$$p_{ij} = Rx_j t_{ji}/x_i.$$

Since T is R- positive, P is positive-recurrent, and so has a unique invariant measure (to constant multiples) $\boldsymbol{\nu}$, which moreover satisfies $\boldsymbol{\nu}'\mathbf{1} < \infty$. (Theorem 5.5 and its periodic version.)

Now consider the row vector $\{x_i y_i\}$; this is in fact an invariant measure for P, since

$$\sum_i x_i y_i p_{ij} = Rx_j \sum_i t_{ji} y_i = x_j y_j$$

for each $j = 1, 2, \ldots$. Hence

$$\sum_j x_j y_j < \infty. \qquad \#$$

THEOREM 6.5. If T is an aperiodic R- positive matrix, then as $k \to \infty$

$$R^k\, t_{ij}^{(k)} \to x_j y_i / \sum_j x_j y_j, > 0,$$

where $\mathbf{x}', \mathbf{y}$ are an invariant measure and vector of T respectively.

Proof. Form the usual positive-recurrent stochastic matrix $P = \{p_{ij}\}$ (which is now aperiodic)

$$p_{ij} = Rx_j t_{ji}/x_i,$$

so that $p_{ij}^{(k)} = R^k x_j t_{ji}^{(k)}/x_i$.

From the body of the proof of the second part of the last theorem

$$R^k x_j t_{ji}^{(k)}/x_i = p_{ij}^{(k)} \to v_j = \frac{x_j y_j}{\sum_j x_j y_j}$$

since $\nu = \{v_j\}$ constitutes the unique invariant measure of P normed to sum to unity (Theorem 5.5). The assertion follows immediately. #

6.3 Consequences for finite and stochastic infinite matrices

(i) *Finite irreducible T*

It has already been mentioned that in this case $1/R = r$, where r is the Perron–Frobenius eigenvalue of T. Clearly, since $T^k/r^k \nrightarrow 0$, such T are R- positive; the unique R- invariant measure and vector are of course the left and right Perron–Frobenius eigenvectors corresponding to the eigenvalue r. A consequence of the new theoretical development is that for each i,

$$L_{ii}(r^{-1}) = 1.$$

Theorem 6.5 is of course a weak version of Theorem 1.2 of Chapter 1. [The question of when the rate of convergence is (uniformly) geometric in the case of infinite T does not have a simple answer – see the discussion following Theorem 6.6 in the next section, and the Bibliography and discussion.]

(ii) *Infinite irreducible stochastic matrices*

For $T = P$, where P is an infinite irreducible stochastic matrix, it has already been noted that $R \geqslant 1$. Viewing the development of those results of §5.3 which relate to the problem of subinvariant measures of such a P, without specialization to recurrent P, such as Theorem 5.3, it is seen without difficulty from Theorem 6.3 that in essence the discussion there pertains in general to β- subinvariance, where $\beta \equiv 1 \leqslant R$, this choice of β having been to some extent dictated by analogy with finite stochastic P, where $1 = 1/r = R$ (although of course invariance itself is of physical significance in Markov chain theory†).

However it is now clear from the earlier section of this chapter that in fact β- subinvariance, with $\beta = 1$, may in general be of less profound significance than R- subinvariance if it may happen that $R > 1$. On the other hand since finite irreducible stochastic P are always R- positive, with $R = 1/r = 1$, we do not expect to be able to improve on the results of Chapter 5 for infinite R- positive T (and perhaps also R- null T) where we would expect $R = 1$.

† See Derman (1955).

THEOREM 6.6. If an irreducible stochastic P is positive-recurrent or null-recurrent, it is R- positive or R- null (respectively) with $R = 1$. Conversely an R- positive or R- null stochastic P with $R = 1$ implies (respectively) positive-recurrence and null-recurrence.

There exist stochastic P which are transient with $R > 1$; such P may still be R- positive.

Proof. If P is recurrent $L_{ii}(1-) = 1$ for any index i, by definition of recurrence. On the other hand $L_{ii}(R-) \leqslant 1$ and for $0 < \beta < R$, clearly

$$L_{ii}(\beta-) < L_{ii}(R-).$$

Hence $R = 1$; and consequently P is recurrent. The rest of the first assertion is trivial, as is its converse.

We demonstrate the final part of the theorem† by the example discussed in §5.6, with

$$p_i = \left\{\frac{1}{2} + \left(\frac{1}{2}\right)^{i+1}\right\} \Big/ \left\{\frac{1}{2} + \left(\frac{1}{2}\right)^{i}\right\}, i \geqslant 1.$$

Then

$$\begin{aligned} l_{11}^{(k)} = f_{11}^{(k)} &= p_0\, p_1 \ldots p_{k-1}\,(1 - p_k), \, (p_0 = 1) \\ &= (1/2)^{k+1}. \end{aligned}$$

Therefore $\quad F_{11}(z) = \sum_{k=1}^{\infty} (1/2)^{k+1} z^k = (z/4)(1 - z/2)^{-1}.$

Hence $F_{11}(1) = 1/2$ so that P is transient.

Moreover $\quad P_{11}(z) = [1 - F_{11}(z)]^{-1} = 2(2 - z)(4 - 3z)^{-1}$

at least for $|z| < 1$. Hence by analytic continuation, the power series $P_{11}(z)$ has convergence radius 4/3 so that

$$R = 4/3 \; (> 1).$$

Also for $0 < s < 4/3$

$$F'_{11}(s) = (2 - s)^{-2}$$

so $$F'_{11}(4/3-) = 9/4 < \infty$$

so that P is R- positive. #

Transient stochastic P with $R > 1$ are sometimes called *geometrically transient* in virtue of the following result.

THEOREM 6.7. A stochastic irreducible P satisfies

$$p_{ij}^{(k)} = 0(\delta^k), \tag{6.3}$$

† See Exercise 6.5 for an example of a transient P which has $R > 1$ and is R- transient.

as $k \to \infty$, for some δ, $0 < \delta < 1$ and for some fixed pair of indices i and j, if and only if P has convergence parameter $R > 1$. In this case (6.3) holds uniformly in δ for each pair i, j, for any fixed δ satisfying $R^{-1} \leqslant \delta < 1$; and for no $\delta < R^{-1}$ for any pair (i, j).

Proof. Consider the power series

$$P_{ij}(z) = \sum_{k=0}^{\infty} p_{ij}^{(k)} z^k.$$

Since $p_{ij}^{(k)} \leqslant \text{const}.\delta^k$, it follows that the power series converges for $|\delta z| < 1$, i.e. $|z| < \delta^{-1}$, and since $\delta^{-1} > 1$ it follows that the common convergence radius R of all such series satisfies $R \geqslant \delta^{-1} > 1$ [and so, by Theorem 6.6, P is transient]. Conversely if P has convergence parameter $R > 1$, then since for any fixed pair i, j

$$p_{ij}^{(k)} R^k$$

is bounded as $k \to \infty$, it follows

$$p_{ij}^{(k)} = 0(R^{-k})$$

as $k \to \infty$. Moreover this is true for any pair of indices i, j. If $\delta < R^{-1}$, $\delta^{-k} p_{ij}^{(k)}$ is unbounded as $k \to \infty$ {since $\beta^k t_{ij}^{(k)}$ is for any $\beta > R$}. #

Note that in general, however, in this situation of geometric transience one can only assert that

$$p_{ij}^{(k)} \leqslant C_{ij} R^{-k}$$

where C_{ij} is a constant which *cannot* be replaced by a uniform constant for all pairs of indices (i, j).

COROLLARY. If P is recurrent it is not possible to find a pair of indices (i, j) and a δ satisfying $0 < \delta < 1$ such that

$$p_{ij}^{(k)} = 0(\delta^k), \text{ as } k \to \infty.$$

{This corollary is of course of chief significance for null-recurrent P.}

The analogous question to these considerations for positive recurrent matrices P (R- positive, with $R = 1$, stochastic P) in case P is also aperiodic, is that of whether and when the convergence:

$$|p_{ij}^{(k)} - \nu_j| \to 0 \tag{6.4}$$

(see Theorems 5.5 and 6.5) where $\boldsymbol{\nu}' = \{\nu_j\}$ is the unique stationary measure of P satisfying $\boldsymbol{\nu}'\mathbf{1} = 1$, takes place at a geometric rate for every pair of indices (i, j). This situation is called *geometric ergodicity*; it is known that if a geometric convergence rate to zero obtains for *one pair of indices* (i, j), the same geometric convergence rate is applicable to all indices, so that geometric ergodicity is a (uniform) *solidarity result* for a positive recurrent aperiodic stochastic set of indices. However it *does not always* hold in such a situation; i.e. positive recurrent aperiodic stochastic P exist for which the rate in (6.4) is geometric for *no pair* of indices (i, j).

6.4 Finite approximations to infinite irreducible T.

Since the theory evolved in this chapter to this point for non-negative T satisfying our basic assumptions coincides with Perron–Frobenius results if T is actually finite, and is thus a natural extension of it to the countable case, a question of some theoretical as well as computational interest pertains to whether, in the case of an infinite T, its Perron–Frobenius-type structure is reflected to some extent in its $(n \times n)$ northwest corner truncations ${}_{(n)}T$, and increasingly so as $n \to \infty$.

To make the investigation easier, it seems natural to make a further assumption for this section, that at least the irreducibility structure of T should be to some extent reflected in the structure of the truncation ${}_{(n)}T$:

Assumption 4: All but at most a finite number of truncations ${}_{(n)}T, n \geqslant 1$, are irreducible.†

We shall adopt the convention that a preceding subscript (n) refers to the truncation ${}_{(n)}T$; thus e.g. (for sufficiently large n) ${}_{(n)}R$ is the convergence parameter of ${}_{(n)}T$, i.e. the Perron–Frobenius eigenvalue of ${}_{(n)}T$ is ${}_{(n)}R^{-1}$.

THEOREM 6.8. ${}_{(n+1)}R < {}_{(n)}R$ for all irreducible ${}_{(n)}T$. ${}_{(n)}R \downarrow R$ as $n \to \infty$.

Proof. Let ${}_{(n)}y = \{{}_{(n)}y_i\}$ be the positive right Perron–Frobenius eigenvector of ${}_{(n)}T$; then the matrix ${}_{(n)}P = \{{}_{(n)}p_{ij}\}$ defined by

$$ {}_{(n)}p_{ij} = {}_{(n)}R \; {}_{(n)}t_{ij} \; {}_{(n)}y_j / {}_{(n)}y_i $$

is stochastic and irreducible, for all n for which ${}_{(n)}T$ is irreducible.

Now the $(n \times n)$ northwest corner truncation of ${}_{(n+1)}P$ must be strictly substochastic, or ${}_{(n+1)}P$ could not be irreducible. On the other hand since this truncation has entries

$$ {}_{(n+1)}R \; {}_{(n+1)}t_{ij} \; {}_{(n+1)}y_j / {}_{(n+1)}y_i $$

$i, j = 1, \ldots, n$, and for these indices ${}_{(n+1)}t_{ij} = {}_{(n)}t_{ij}$, it follows that it has positive entries in the same position as ${}_{(n)}P$ and so is irreducible; consequently *its* Perron–Frobenius eigenvalue is *strictly less* than unity. On the other hand its convergence radius is clearly ${}_{(n)}R / {}_{(n+1)}R$, since ${}_{(n+1)}t_{ij} = {}_{(n)}t_{ij}$, $i, j = 1, 2, \ldots, n$.

Therefore $\qquad {}_{(n)}R / {}_{(n+1)}R > 1$, as required.

† This assumption does not in fact result in essential loss of generality – see the Bibliography and discussion to this chapter.

Thus ${}_{(n)}R \downarrow {}_{(\infty)}R$ for some ${}_{(\infty)}R$ satisfying $R \leqslant {}_{(\infty)}R < {}_{(n)}R$, since ${}_{(n)}t_{11}^{(k)} \leqslant t_{11}^{(k)}$ implies that $R \leqslant {}_{(n)}R$.

Now, we have

$${}_{(n)}R \sum_{j=1}^{n} {}_{(n)}t_{ij}\, {}_{(n)}y_j = {}_{(n)}y_i$$

where we assume that for every n (sufficiently large) ${}_{(n)}y_i$ has been scaled so that ${}_{(n)}y_1 = 1$. Put

$$y_j^* = \liminf_{n\to\infty} {}_{(n)}y_j$$

for each $j = 1, 2, \ldots$, so that $y_1^* = 1, \infty \geqslant y_j^* \geqslant 0$. By Fatou's Lemma

$${}_{(\infty)}R \sum_{j=1}^{\infty} t_{ij}\, y_j^* \leqslant y_i^*,\ i = 1, 2, \ldots$$

Iterating this, we find

$${}_{(\infty)}R^k \sum_{j=1}^{\infty} t_{ij}^{(k)}\, y_j^* \leqslant y_i^*$$

and taking $i = 1$, using the fact that $y_1^* = 1$ and the irreducibility of T, implies $y_j^* < \infty$ for all j. Hence $\boldsymbol{y}^* = \{y_i^*\}$ is an ${}_{(\infty)}R$- subinvariant vector; hence $\boldsymbol{y}^* > \boldsymbol{0}$, and ${}_{(\infty)}R \leqslant R$, the last by analogy with part (d) of Theorem 6.3.

Hence finally

$${}_{(\infty)}R = R$$

which is as required. #

THEOREM 6.9. Let ${}_{(n)}\boldsymbol{x}' = \{{}_{(n)}x_i\}$ and ${}_{(n)}\boldsymbol{y} = \{{}_{(n)}y_i\}$ be the left and right Perron–Frobenius eigenvectors, normed so that ${}_{(n)}x_1 = 1 = {}_{(n)}y_1$, of ${}_{(n)}T$, where the infinite matrix T is R- recurrent. Then

$$\lim_{n\to\infty} {}_{(n)}y_i = y_i, \lim_{n\to\infty} {}_{(n)}x_i = x_i$$

exist, for each i, and $\boldsymbol{x}' = \{x_i\}$ and $\boldsymbol{y} = \{y_i\}$ are the unique R- invariant measure and vector, respectively, of T, normed so that $x_1 = 1 = y_1$.

Proof. We give this for the $\boldsymbol{y}$ only, as is adequate. From the proof of Theorem 6.8

$$\boldsymbol{y}^* = \{y_i^*\}, y_i^* = \liminf_{n\to\infty} {}_{(n)}y_i$$

is always an R- subinvariant vector of T, and when T is R- recurrent (as at present) it must be the unique R- invariant vector of T with first element unity, $\boldsymbol{y}$. (Theorem 6.2.)

Let i^* be the first index for which

$$y_{i^*}^* = \liminf_{n\to\infty} {}_{(n)}y_{i^*} < \limsup_{n\to\infty} {}_{(n)}y_{i^*} \leqslant \infty$$

Then there exists a subsequence $\{n_r\}$ of the integers such that

$$\overline{y}_{i*} \equiv \lim_{r\to\infty} {}_{(n_r)}y_{i*} = \limsup_{n\to\infty} {}_{(n)}y_{i*} \leqslant \infty.$$

If we repeat the relevant part of the argument in the proof of Theorem 6.8, as in the proof of this theorem, but using the subsequence $\{n_r\}, r = 1, 2, \ldots$ rather than that of all integers, we shall construct an R- invariant vector of T, $\overline{y} = \{\overline{y}_i\}$ with first element unity, but with

$$\overline{y}_{i*} > y_{i*}^* \equiv y_{i*}$$

which is impossible, by uniqueness of the R- invariant vector $\boldsymbol{y}$ with first element unity. #

In the following theory we concentrate on how the determinantal and cofactor properties of the matrices $[{}_{(n)}I - R_{(n)}T]$ relate to those of the infinite matrix $[I - RT]$. Since ${}_{(n)}R > R$, the reader will recognize here that there is a relationship between this theory and that of §2.1 of Chapter 2. In actual fact we shall not pursue the relation to any great extent, persisting rather in the investigation of the approximative properties of the finite truncations to the Perron–Frobenius-type structure of T. Some investigation of the determinantal properties of $[{}_{(n)}I - \beta_{(n)}T]$ as $n \to \infty$, for $\beta \leqslant R$ however occurs in the Exercises.†

Write,‡

$${}_{(n)}\Delta(\beta) = \det\,[{}_{(n)}I - \beta \cdot {}_{(n)}T]\,;\; {}_{(n)}\Delta \equiv {}_{(n)}\Delta(R).$$

$${}_{(n)}c_{ij}(\beta) = \text{cofactor of the } (i, j) \text{ entry of } [{}_{(n)}I - \beta \cdot {}_{(n)}T]\,;\; {}_{(n)}c_{ij} \equiv {}_{(n)}c_{ij}(R).$$

We note that for $0 \leqslant \beta \leqslant R$,

$${}_{(n)}\Delta(\beta) > 0$$

since $\beta < {}_{(n)}R$ [this follows from a fact already used in the proof of Theorem 1.1, that for a real square matrix H, the characteristic function $\det[\lambda I - H]$ is positive for λ exceeding the largest real root, if one such exists].

We shall now consider only the quantities ${}_{(n)}\Delta$, $\{{}_{(n)}c_{ij}\}$.

THEOREM 6.10. (*a*) As $n \to \infty$,

$${}_{(n)}c_{ji}/{}_{(n)}c_{ii} \uparrow \overline{x}_{ij} > 0 \; (i, j = 1, 2, \ldots),$$

where, for fixed i, $\{\overline{x}_{ij}\}$, $j = 1, 2, \ldots$ is the minimal left R- subinvariant measure¶ of T.

† See Exercise 6.8.

‡ Noting that the definition of these symbols differs here from that in Chapter 2, slightly.

¶ As defined at the outset of §6.2. A similar proposition will, of course, hold for R- subinvariant vectors.

(*b*) $\lim_{n\to\infty} {}_{(n)}\Delta \geqslant 0$ exists; if $\lim {}_{(n)}\Delta > 0$, T is R- transient. $\lim_{n\to\infty} {}_{(n)}c_{ij} \geqslant 0$ exists for every pair (i, j), and all these limits are positive or zero together. For an R- transient matrix, ${}_{(n)}\Delta$ and ${}_{(n)}c_{ij}$ have positive or zero limits together.

Proof. Since $R_n > R$, and ${}_{(n)}c_{ji}/{}_{(n)}\Delta$ is the (i, j) entry of $[{}_{(n)}I - R_{(n)}T]^{-1}$, it follows from Lemma B.1 of Appendix B that

$$ {}_{(n)}c_{ji} = {}_{(n)}\Delta \,.\, {}_{(n)}T_{ij}(R), \text{ so that } {}_{(n)}c_{ji} > 0 \text{ for } i, j = 1, 2, \ldots $$

Further, when $i \neq j$, since for $|z| < {}_{(n)}R$, ${}_{(n)}T_{ij}(z) = {}_{(n)}L_{ij}(z)\,(1 - {}_{(n)}L_{ii}(z))^{-1}$ it follows that

$$ {}_{(n)}L_{ij}(R) \cdot {}_{(n)}\Delta = {}_{(n)}c_{ji}\,[1 - {}_{(n)}L_{ii}(R)]\,,\; i \neq j $$

$$ {}_{(n)}\Delta = {}_{(n)}c_{ii}\,[1 - {}_{(n)}L_{ii}(R)]\,. $$

Hence for $i \neq j$,

$$ {}_{(n)}c_{ji}/{}_{(n)}c_{ii} = {}_{(n)}\,L_{ij}(R) $$

and since it is clear that ${}_{(n)}\ell_{ij}^{(k)} \uparrow \ell_{ij}^{(k)}$ as $n \to \infty$,

$$ {}_{(n)}c_{ji}/{}_{(n)}c_{ii} \uparrow L_{ij}(R)\;(<\infty) $$

as $n \to \infty$, which proves (*a*). [The inequality and the rest of the assertion follows from Theorems 6.2 and 6.3.]

We now pass to the proof of (*b*). By cofactor expansion along the $(n + 1)$th column of: $[{}_{(n+1)}I - R\,.\,{}_{(n+1)}T]$ *with its jth row and column deleted,*

$$ {}_{(n+1)}c_{jj} = (1 - R \cdot t_{n+1,n+1})\,{}_{(n+1)}c_{jj/n+1,n+1} - R \sum_{\substack{k=1\\k\neq j}}^{n} t_{k,n+1}\,.\,{}_{(n+1)}c_{jj/k,n+1} $$

$$ = {}_{(n+1)}c_{jj/n+1,n+1} - R \sum_{\substack{k=1\\k\neq j}}^{n+1} t_{k,n+1} \cdot {}_{(n+1)}c_{jj/k,n+1}\,; $$

and,

$$ {}_{(n+1)}c_{jj/n+1,n+1} = {}_{(n)}c_{jj}; $$

where ${}_{(n+1)}c_{jj/k,n+1}$ is the $(k,\, n + 1)$ cofactor of this matrix. The corresponding matrix formed from ${}_{(n+1)}T$ by crossing out its jth row and column, may not be irreducible, but none of its eigenvalues can exceed $1/{}_{(n+1)}R$ in modulus†. Hence

$$ {}_{(n+1)}c_{jj/k,n+1} \geqslant 0 $$

in the same manner as in the first step of the proof of this theorem. Thus

$$ 0 < {}_{(n+1)}c_{jj} \leqslant {}_{(n)}c_{jj}, $$

so that

$$ \lim_{n\to\infty} {}_{(n)}c_{jj} \geqslant 0 $$

† See e.g. Exercise 1.12, (*e*′).

exists; let us put c_{jj} for the value of this limit. Thus as $n \to \infty$

$$ {}_{(n)}\Delta \downarrow \lim_{n\to\infty} {}_{(n)}c_{ii}[1 - {}_{(n)}L_{ii}(R)] $$

$$ = c_{ii}[1 - L_{ii}(R)] \geqslant 0 $$

exists; and if the limit is positive $L_{ii}(R) < 1$, so T is R- transient. Further, since

$$ {}_{(n)}c_{ji}/{}_{(n)}c_{ii} \to \bar{x}_{ij} > 0 $$

it follows that if for some i, $c_{ii} > 0$ then the limit

$$ c_{ji} = \lim_{n\to\infty} {}_{(n)}c_{ji} $$

exists and $c_{ji} > 0$, for all $j = 1, 2, \ldots$ and if $c_{ii} = 0$,

$$ 0 = c_{ji} = \lim_{n\to\infty} {}_{(n)}c_{ji} $$

exists. Thus we have that in the matrix whose (i, j) entry is c_{ji} each row is either strictly positive or zero.

By making the same considerations as hitherto for the *transpose* of T, T', we shall obtain the same conclusion about the columns of this same matrix, whose (i, j) entry is c_{ji}; so that finally *all* c_{ji} are positive or zero as asserted.

Finally, if T is R- transient, $L_{ii}(R) < 1$, and since

$$ \lim {}_{(n)}\Delta = c_{ii}[1 - L_{ii}(R)] $$

the conclusion that $\lim_{(n)}\Delta$ and the c_{ij} are positive or zero together, follows. #

COROLLARY. For any pair (i, j)

$$ \frac{{}_{(n)}c_{ji}}{{}_{(n)}c_{ii}} \leqslant \bar{x}_{ij} \leqslant \left\{\bar{x}_{ji}\right\}^{-1} \leqslant \left\{\frac{{}_{(n)}c_{ij}}{{}_{(n)}c_{jj}}\right\}^{-1}; $$

with both sides converging to $\bar{x}_{ij} = \{\bar{x}_{ji}\}^{-1}$ as $n \to \infty$, if T is R- recurrent.

Proof. The inequalities

$$ \frac{{}_{(n)}c_{ji}}{{}_{(n)}c_{ii}} \leqslant \bar{x}_{ij}; \left\{\bar{x}_{ji}\right\}^{-1} \leqslant \left\{\frac{{}_{(n)}c_{ij}}{{}_{(n)}c_{jj}}\right\}^{-1}, $$

are given by the theorem, as is the convergence in each. Let $\{x(i)\}$ be any R- subinvariant measure of T. Then by the minimality property of $\{\bar{x}_{ij}\}$

$$ \bar{x}_{ij} \leqslant x(j)/x(i) = \{x(i)/x(j)\}^{-1} \leqslant \{\bar{x}_{ji}\}^{-1} \tag{6.5} $$

which completes the proof of the inequality.

If T is R- recurrent, $\{\bar{x}_{ij}\}$, $j = 1, 2, \ldots$ is the only R- invariant measure of T with ith element unity. Hence all inequalities in (6.5) become equalities, which completes the assertion. #

Thus we can compute precise bounds, from ${}_{(n)}T$ for the unique R- invariant measure with ith element unity of an R- recurrent T.

Truncations of infinite stochastic matrices

For irreducible infinite stochastic matrices P satisfying the additional assumption of irreducible truncations, it is of some interest, in view of §6.3, to glance at the consequences of evolving the above theory with $\beta = 1$, rather than $\beta = R$, in general, although we know now, in view of Theorem 6.6, that it is only for transient P that there may be differing results.

On the other hand, even with recurrent P, the stochasticity of P will imply a stronger structure, than in general, for e.g. the c_{ij} if all are positive.

THEOREM 6.11. If $T = P$ is stochastic,

(*a*) c_{ji} is independent of i, if P is recurrent;

(*b*) $\lim_{n\to\infty} {}_{(n)}\Delta(1)$ exists and is non-negative. A positive limit implies transience of P, and positivity of all $c_{ij}(1)$.

(*c*) There exist recurrent P where all c_{ij} are positive; and transient P where $\lim_{n\to\infty} {}_{(n)}\Delta(1) > 0$.

Proof. We need to prove (*a*) only in the case where all c_{ij} are positive. Supposing we follow the steps of the previous theorem with the quantities ${}_{(n)}F_{ij}(R)$ rather than the ${}_{(n)}L_{ij}(R)$ – an approach which would be more appropriate to the consideration of R- subinvariant vectors – we would find eventually that

$$c_{ji} = c_{jj}\, F_{ij}(R\text{-}) = c_{jj}\, F_{ij}(1\text{-})$$

since in this case $R = 1$; and so

$$c_{ji} = c_{jj}$$

since F_{ij} (1-) = 1 for all i, j for a recurrent P (Theorem 5.4^{D}).

To prove (*b*), we note that, in general, since ${}_{(n)}R > R \geqslant 1$, the proof can be carried forward as in that of Theorem 6.10, but with 1 replacing R, to obtain

$$\lim_{n\to\infty} {}_{(n)}\Delta(1) = c_{ii}(1)\ [1 - L_{ii}(1\text{-})] \geqslant 0,$$

where $c_{ii}(1)$ is defined as $\lim_{n\to\infty} {}_{(n)}c_{ii}(1)$.

To demonstrate the validity of assertion (*c*), we again return to the stochastic P investigated in §5.6, which we note has all its truncations ${}_{(n)}P$ irreducible. Moreover,

$$\begin{aligned} {}_{(n)}\Delta(1) &= \det[{}_{(n)}I - {}_{(n)}P] \\ &= (1 - q_n)\, p_1 p_2 \ldots\ p_{n-1}, = \alpha_n \end{aligned}$$

in the notation of §5.6, the determinant being evaluated by first adding each column in turn to the first; and then expanding by the nth row.

On the other hand for $1 \leqslant j \leqslant n$

$$ {}_{(n)}c_{jj} = p_0 p_1 \ldots p_{j-1}, (p_0 = 1) $$
$$ = \alpha_{j-1} $$

by deleting the jth row and column of $[{}_{(n)}I - {}_{(n)}P]$, adding all columns in turn to the first in the resulting matrix; and finally expanding the determinant by its first column.

Thus for $j = 1, 2, \ldots$

$$ c_{jj} = \alpha_{j-1} $$

irrespective of whether the matrix P is recurrent or transient. On the other hand P is transient if and only if $\alpha_\infty > 0$ and so (in this case) P is transient if and only if

$$ \alpha_\infty \equiv \lim_{n\to\infty} {}_{(n)}\Delta(1) > 0. \quad \# $$

As two final comments on the stochastic situation, firstly if P is in fact *positive-recurrent*, we are in general interested in its unique invariant measure $\boldsymbol{x}'$ normed *probabilistically*, i.e. so that $\boldsymbol{x}'\mathbf{1} = 1$, whereas both the algorithms here presented (in Theorems 6.9 and 6.10) would yield the unique invariant measure normed so that a specified element is unity.

The problem of constructing a truncation algorithm for positive-recurrent P where approximative quantities converge elementwise to the probabilistically normed invariant measure (stationary distribution) has been solved only under additional restrictions.

The computationally important problem of the rates of convergence of the approximations to their limit values as $n \to \infty$ is also difficult.

Some indications of work done in these directions is given in the next section.

Secondly, on the theoretical side there may be probabilistic grounds for calling those transient P where $\lim {}_{(n)}\Delta(1) > 0$, *strongly transient* since in the example considered above and an appropriate generalization of it,† the positivity of this limit indicates that there is a *positive probability* of a corresponding Markov chain 'going to infinity' by the *'shortest possible'* route (in these cases well defined) no matter what the initial state-index, the limit itself being this probability if the chain starts at state-index 1.

Bibliography and discussion

The strikingly elegant extension of Perron–Frobenius theory to infinite irreducible non-negative matrices T is as presented in §6.1–§6.3 basically due to D. Vere-Jones (1962, 1967) where a more detailed treatment and further results may be found. The reader is referred also to Albert (1963), Pruitt (1964), Kendall (1966), and Moy (1967*a*) in this connection; the last reference

† See Exercise 6.9.

gives a particularly detailed treatment of the periodic case of T (see also Šidák (1964*a*)) and a generalization to irreducible T of the Martin boundary and potential theory of §5.5. The latter part of the proof of Theorem 6.1 is due to Kingman (1963, §8), and Theorem 6.4 is due to Kendall (1966).

As regards the Basic Assumptions 1–3 on T, consequences of relaxing 2 (i.e. of not necessarily assuming irreducibility) are examined briefly by Kingman (1963, §6) and Tweedie (1971, §2). The role of 1 and 2 and the consequences of their relaxation are examined by Kendall (1966) and Mandl & Seneta (1969; §2). Mandl & Seneta (1969) also extend the theory of §3.1 to countable T.

For infinite irreducible stochastic P the study of problems of geometric rate of convergence to their limit was initiated by Kendall (1959) and it is to him that the term *geometric ergodicity,* and the proof of its being a solidarity property, are due. The study was continued by Kendall (1960); Vere-Jones (1962) proved the uniform solidarity results for geometric convergence rate for transient and positive-recurrent P discussed in Theorem 6.7 and its following remarks: and further contributions were made by Vere-Jones (1963) Kingman (1963) and Vere-Jones (1966).

For examples of the usage of the Vere-Jones R- theory see also Seneta & Vere-Jones (1966), Moy (1967*b*), Daley (1969) and Kuich (1970*a*). Study of countable non-negative T as operators on sequence spaces has been undertaken by Putnam (1958, 1961) and, comprehensively, by Vere-Jones (1968).

The contents of §6.4 are taken from Seneta (1967*a*), a paper largely motivated by some earlier results and ideas of Sarymsakov (1953*b*; 1954, §§22–24) for infinite primitive stochastic P. Thus the result ${}_{(n)}R \downarrow {}_{(\infty)}R \geqslant 1$ is due to him, as is the idea of the proof of Theorem 6.10; although in the absence at the time of Vere-Jones' extension of the Perron–Frobenius theory to infinite irreducible T, his results were necessarily rather weaker. The present proof of Theorem 6.8 is adapted from Mandl & Seneta (1969) and differs from the proof in Seneta (1967*a*). The Corollary to Theorem 6.10 is essentially due to Kendall (see Seneta, 1967*a*).

A discussion of the fact that Assumption 4 in §6.4 does not result in essential loss of generality is given in Seneta (1968*a*) as a result of correspondence between Kendall and Seneta; and to some extent in Sarymsakov (1953*b*, pp. 11–12). Some computational aspects of the two algorithms of the theory are briefly discussed in this 1968 paper also, and analysed, extended and implemented in Allen, Anderssen & Seneta (1973).

A theoretical discussion of the truncation theory when the irreducibility condition on T is relaxed, is given by Tweedie (1971).

The problem of convergence to the unique stationary distribution of a denumerable stochastic P which is a *Markov matrix,* i.e. satisfies

$$\sup_j \left(\inf_i p_{ij} \right) > 0$$

has been solved by Golub & Seneta (1973).

Other references to the use of truncated matrices in various probabilistic contexts are given in Seneta (1967*a*). The connection between the classical theory of infinite determinants and that of the existence and positivity of $\lim_{n\to\infty} {}_{(n)}\Delta(1)$ in the case of stochastic P is sketched by Seneta (1968*b*).

Other examples of the effects of the truncation method on substochastic and stochastic matrices are given by Sarymsakov (1954, §24) and Seneta (1967*a*); and in Exercises 6.7, 6.8 and 6.9.

Exercises

6.1. Prove the last part of Theorem 6.1 (avoiding the use of Lemma A.4) by assuming to the contrary that $R = \infty$ and using the relation $T_{ii}(z) = (1 - L_{ii}(z))^{-1}$.

Hint: For some K, $l_{ii}^{(k)} > 0$.

6.2. Show that if the Basic Assumption 3 is not made, then the situation where the convergence parameter R is 0 must be defined as R- transient to accord with the present theory for $R > 0$.

(Kendall, 1966)

6.3. Show that if T is periodic, with period d, then there exist 'eigenvectors' (with possibly complex elements) corresponding to each of the eigenvalues $R^{-1} e^{2\pi i h/d}$, $h = 0, 1, 2, \ldots, d-1$ if T possesses an R- invariant vector.

6.4. Suppose A and B are finite or infinite irreducible non-negative matrices, and R_A, $R_B \geq 0$ their convergence parameters. Suppose X is a non-negative, non-zero matrix such that AX, XB are elementwise finite, and that

$$AX = XB.$$

Show that, if each row of X has only a finite number of non-zero elements, $R_B \leq R_A$; and if each column of X has only a finite number of non-zero elements $R_A \leq R_B$. (Thus if X is *both row-finite* and *column-finite* in the above sense, $R_A = R_B$.)

Hint: Show first $A^k X = XB^k$.

(Generalized from Kuich, 1970*b*)

6.5. A semi-unrestricted random walk with reflecting barrier at the origin, as described by the stochastic matrix P of Example (3) of Chapter 4, §4.1, with $a = q$ can be shown to have

$$F_{00}(z) = qz + (1/2)\,(1 - \sqrt{[1 - 4pqz^2]}).$$

(Recall that in this instance the index set is $\{0, 1, 2, \ldots\}$.)

Show that P is transient if and only if $p > q$; in which case $R = (4pq)^{-1/2}$ and P is R- transient.

Hint: $1 - 4pq = (p + q)^2 - 4pq$.

6.6. Let $A = \{a_{ij}\}$ be an infinite matrix defined by $a_{i,i} = 0$, $a_{i,2i} = c_1$, $a_{i,2i+1} = c_2$, $a_{2i,i} = d$, $a_{2i+1,i} = d$ $(i \geq 1)$, $a_{i,j} = 0$ otherwise; c_1, c_2, $d > 0$. Show A is irreducible.

Construct a non-negative matrix X which is row and column finite such that $AX = XB$ where $B = \{b_{ij}\}$, is given by

$$b_{i,i+1} = c_1 + c_2,\ i \geqslant 1;\ b_{i,i-1} = d,\ i \geqslant 2,$$

all other $b_{ij} = 0$.

For the case $c_1 + c_2 = p$, $0 < p < 1$, it can be shown that for the matrix B,

$$F_{11}(z) = (1/2)\,(1 - \sqrt{[1 - 4pqz^2]}).$$

Use this fact and the result of Exercise 6.5 to deduce that

$$R_A = \{4d(c_1 + c_2)\,(d + c_1 + c_2)\}^{-1/2}.$$

Hint: Take $X = \{x_{ij}\}$ to be such that its jth column contains only zeros, apart from a 'block' of elements with values 1, of length 2^{j-1}.

(Kuich, 1970*b*)

6.7. Show that if T has period d, and all the assumptions of §6.4 are satisfied, all ${}_{(n)}T$ for sufficiently large n have period d also.

(Seneta, 1967*a*)

6.8. Carry through as far as possible the arguments of Theorem 6.10 in the case of the more general matrix $[I - \beta T]$, $0 < \beta \leqslant R$. Show in particular that

$$d(\beta) \overset{def}{=} \lim_{n\to\infty} \det[{}_{(n)}I - \beta \cdot {}_{(n)}T] \geqslant 0$$

exists for $0 \leqslant \beta \leqslant R$ [the approach to the limit being monotone decreasing as n increases]. Thus $d(\beta)$ may, in a sense be regarded as a (modified) characteristic polynomial of T; note that $d(R) = 0$ certainly for R- recurrent T, as one might expect.

6.9. 'Slowly spreading' infinite stochastic matrices $P = \{p_{ij}\}$, $i, j \geqslant 1$ are defined by the condition

$$p_{i,i+r} = 0,\ i \geqslant 1,\ r \geqslant 2$$

(i.e. all elements above the superdiagonal are zero). Irreducibility, which we assume, clearly implies $p_{i,i+1} > 0$, $i \geqslant 1$: this situation clearly subsumes both the Example of §5.6, used in the proof of Theorem 6.11, and that of Exercise 6.5 above.

Show that

$$\Delta_n(1) = \det[{}_{(n)}I - {}_{(n)}P] = \sum_{i=1}^{n} p_{i,i+1}$$

so that

$$\lim_{n\to\infty} \Delta_n(1) > 0 \Leftrightarrow \sum_{i=1}^{\infty} (1 - p_{i,i+1}) < \infty.$$

Note that this condition is far from satisfied in the situation of Exercise 6.5, even if $p > q$ i.e. $p > \frac{1}{2}$ (after adjusting the index set in that question appropriately).

(Adapted from Kemeny, 1966)

6.10. For the matrix $P = \{p_{ij}\}$, $i, j \geqslant 1$, defined by

$$p_{i,i+1} = p_i, \; i \geqslant 1$$

$$p_{i,i-1} = q_i = 1 - p_i, \; i \geqslant 2$$

$$p_{1,1} = q_1$$

where $0 < p_i < 1$ for all $i \geqslant 1$, show that an invariant measure $\boldsymbol{x}' = \{x(i)\}$ always exists and is unique to constant multiples by solving the difference equation

$$p_{i-1}\, x(i-1) + q_{i+1}\, x(i+1) = x(i), \; \mathrm{i} \geqslant 1$$

under the auxiliary condition $x(2) = (p_1/q_2)\, x\,(1)$, where $x(1)$ is taken as positive but otherwise arbitrary.

Hence use the result of Exercise 6.9 to show that there exist stochastic P for which an invariant measure exists, even though

$$\lim_{n\to\infty} {}_{(n)}\Delta(1) > 0$$

(again, contrary to expectations, but again only in the transient case).

APPENDICES

Appendix A. Some elementary number theory

We consider the set of all integers, both non-negative and negative.

LEMMA A.1. Any subset, S, of the integers containing at least one non-zero element and closed under addition and subtraction contains a least positive element and consists of all multiples of this integer.

Proof. Let $a \in S$, $a \neq 0$. Then S contains the difference $a - a = 0$, and also $0 - a = -a$. Consequently there is at least one positive element, $|a|$, in S, and hence there is a smallest positive element, b, in S. Now, S must contain all integral multiples of b, for if it contains nb, $n = 1, 2, \ldots$ etc., then it must contain $(n + 1)b = nb + b$, and we know it contains b. Moreover $(-n)b = 0 - (nb)$ is the difference of two elements in S, for $n = 1, 2, \ldots$ and so S contains all negative multiples of b also.

We now show that S can contain nothing but integral multiples of b. For if c is any element of S, there exist integers q and r such that $c = bq + r$, $0 \leqslant r < b$ (qb is the multiple of b closest to c from below; we say $c \equiv r \pmod{b}$). Thus $r = c - bq$ must also be in S, since it is a difference of numbers in S. Since $r \in S$, $r \geqslant 0$, $r < b$, and b is the least positive integer in S, it follows $r = 0$, so that $c = qb$. #

DEFINITION A.1. Every positive integer which divides all the integers $a_1, a_2, \ldots, a_k$ is said to be a *common divisor* of them. The largest of these common divisors, is said to be the *greatest common divisor* (g.c.d.). This number is a well defined positive number if not all $a_1, a_2, \ldots, a_k$ are zero, which we assume henceforth.

LEMMA A.2. The greatest common divisor of $a_1, a_2, \ldots, a_k$, say d, can be expressed as a 'linear combination', with integral coefficients, of $a_1, a_2, \ldots, a_k$, i.e.

$$d = \sum_{i=1}^{k} b_i a_i, \ b_i \text{ integers.}$$

Proof. Consider the set S of all numbers of the form $\sum_{i=1}^{k} b_i a_i$. For any two such

$$\sum_{i=1}^{k} b_i^{(1)} a_i \pm \sum_{i=1}^{k} b_i^{(2)} a_i = \sum_{i=1}^{k} (b_i^{(1)} \pm b_i^{(2)})\, a_i,$$

and hence the set S of such numbers is closed under addition and subtraction, hence by Lemma A.1 consists of all multiples of some minimum positive number

$$\nu = \sum_{i=1}^{k} b_i a_i.$$

Thus d, the greatest common divisor of $a_1, \ldots, a_k$, must divide ν, so that $0 < d \leqslant \nu$. Now each a_i is itself a member of S (choose the bs so that $b_i = 1$ and all other bs zero), so that each a_i is a multiple of ν, by Lemma A.1. Thus a contradiction arises unless $d = \nu$, since d is supposed to be the greatest common divisor of the as. #

DEFINITION A.2. Let a_i, $i = 1, 2, \ldots$ be an infinite set of positive integers. If d_k is the greatest common divisor of $a_1, \ldots, a_k$, then the greatest common divisor of a_i, $i = 1, 2, \ldots$, is defined by

$$d = \lim_{k \to \infty} d_k.$$

The limit $d \geqslant 1$ clearly exists (since the sequence $\{d_k\}$ is non-increasing).

Moreover d is an integer, and must be attained after a finite number of k, since all the d_ks are integers.

LEMMA A.3. An infinite set of positive integers, $V = \{a_i, i \geqslant 1\}$, which is closed under addition (i.e. if two numbers are in the set, so is their sum), contains all but a finite number of positive multiples of its greatest common divisor.

Proof. We can first divide all elements in the infinite set V by the greatest common divisor d, and thus reduce the problem to the case $d = 1$, which we consider henceforth.

By the fact that $d = 1$ must be the greatest common divisor of some finite subset, $a_1, a_2, \ldots, a_k$, of V, it follows from Lemma A.2 that there is a linear combination of these as such that

$$\sum_{i=1}^{k} a_i n_i = 1,$$

where the n_is are integers. Let us rewrite this as

$$m - n = 1 \qquad \text{(A.1)}$$

where m is the sum of the positive terms, and $-n$ the sum of the negative terms. Clearly both n and m are positive integer linear combinations of the a_is and so belong to the set V. Now, let q be any integer satisfying $q \geqslant n \times (n-1)$; and write

$$q = an + b,\ 0 \leqslant b < n$$

where a is a positive integer, $a \geqslant (n-1)$. Then using (A.1)

$$q = an + b(m-n)$$

$$q = (a-b)n + bm$$

so that q is also in the set V. Hence all sufficiently large integers are in the set V, as required. #

COROLLARY. If $a_1, a_2, \ldots, a_k$ are positive integers with g.c.d. unity, then any sufficiently large positive integer q may be expressed as

$$q = \sum_{i=1}^{k} a_i p_i$$

where the p_i are non-negative integers.

We conclude this subsection by an application which is of relevance in Chapter 6, and whose proof involves, in a sense, a tightening of the kind of argument given in Lemma A.3.

LEMMA A.4.† Let u_i $(i = 0, 1, 2, \ldots)$ be non-negative numbers such that, for all $i, j \geqslant 0$

$$u_{i+j} \geqslant u_i u_j.$$

Suppose the set V of those integers $i \geqslant 1$ for which $u_i > 0$ is non-empty and has g.c.d., say d, which satisfies $d = 1$. Then

$$u = \lim_{n \to \infty} u_n^{1/n}$$

exists and satisfies $0 < u \leqslant \infty$; further, for all $i \geqslant 0$, $u_i \leqslant u^i$.

Proof. The set V is closed under addition, in virtue of $u_{i+j} \geqslant u_i u_j$, and since $d = 1$, by Lemma A.3, V contains all sufficiently large integers. Hence for any $r \in V$ there exists an $s \in V$, $s > r$, such that the g.c.d. of r and s is unity. Thus by Lemma A.2 we have

$$1 = b_1 r + b_2 s$$

for some integers b_1, b_2. Not both of b_1 and b_2 can be strictly positive, since $r, s \geqslant 1$. Assume for the moment

$$b_1 \equiv a > 0, -b_2 \equiv b \geqslant 0.$$

Let n be any integer such that $n \geqslant rs$, and let $k = k(n)$ be the smallest positive integer such that

$$b/r \leqslant k/n \leqslant a/s;$$

† Due to Kingman (1963). This is an analogue of the usual theorems on supermultiplicative or subadditive functions (e.g. Hille & Phillips 1957, Theorem 7.6.1; Khintchine, 1969, § 7)

such an integer certainly exists since

$$n(a/s - b/r) = n(ar - bs)/rs = n/rs \geqslant 1.$$

Then
$$n = Ar + Bs$$

where $A = na - k(n)s$, $B = k(n)r - nb$ are non-negative integers. Now, from the assumption of the Lemma,

$$u_n = u_{Ar+Bs} \geqslant u_{Ar}\, u_{Bs} \geqslant \ldots \geqslant u_r^A\, u_s^B, \qquad (n \geqslant rs).$$

Thus
$$u_n^{1/n} \geqslant u_r^{a-(ks/n)}\, u_s^{(kr/n)-b}.$$

Letting $n \to \infty$, and noting $k(n)/n \to b/r$, we see that

$$\liminf_{n\to\infty} u_n^{1/n} \geqslant u_r^{a-bs/r}\, u_s^{b-b} = u_r^{1/r}.$$

This holds for any $r \in V$ and so for any sufficiently large integer r. Hence

$$\liminf_{n\to\infty} u_n^{1/n} \geqslant \limsup_{r\to\infty} u_r^{1/r}$$

which shows
$$u = \lim_{n\to\infty} u_n^{1/n}$$

exists. Further, again from the inequality $u_{i+j} \geqslant u_i u_j$,

$$u_{ki} \geqslant u_i^k,\ i,\ k \geqslant 0$$

so that
$$(u_{ki}^{1/ki})^i \geqslant u_i$$

and letting $k \to \infty$,

$$u^i \geqslant u_i,\ i \geqslant 1$$

and since $u_i > 0$ for some $i \geqslant 1$ the proof is complete, apart from the case $i = 0$, which asserts

$$1 \geqslant u_0.$$

The truth of this follows trivially by putting $i = j = 0$ in the inequality, to obtain $u_0 \geqslant u_0^2$.

We need now to return to the possibility that $b_1 \leqslant 0, b_2 > 0$. If we write $b_2 = a$, $-b_1 = b$, so that $1 = as - br$, then the roles of s and r need to be changed in the subsequent argument and, apart from this $k = k(n)$ needs to be chosen as the *largest* integer such that

$$b/s \leqslant k/n \leqslant a/r.$$

This leads eventually once more to

$$\liminf_{n\to\infty} u_n^{1/n} \geqslant u_r^{1/r}$$

and the rest is as before. #

COROLLARY

$$u = \sup_n u_n^{1/n}.$$

Appendix B. Some general matrix lemmas

LEMMA B.1. If A is a finite $n \times n$ matrix with real or complex elements such that $A^k \to 0$ elementwise as $k \to \infty$, then $(I - A)^{-1}$ exists and

$$(I - A)^{-1} = \sum_{k=0}^{\infty} A^k$$

convergence being elementwise. ($A^0 = I$ by definition.)

Proof. First note that

$$(I - A)(I + A + \ldots + A^{k-1}) = I - A^k.$$

Now, for k sufficiently large, A^k is uniformly close to the zero matrix, and so $I - A^k$ is to I, and is therefore non-singular. (More specifically, by the continuity of the eigenvalues of a matrix if its elements are perturbed, the eigenvalues of $I - A^k$ must be close to those of I for large k, the latter being all 1; hence $I - A^k$ has no zero eigenvalues, and is therefore non-singular.) Taking determinants

$$\det (I - A) \det (I + A + \ldots + A^{k-1}) = \det (I - A^k) \neq 0,$$

therefore $$\det (I - A) \neq 0$$

Therefore

$$(I - A)^{-1} \text{ exists, and } I + A + \ldots + A^{k-1} = (I - A)^{-1} (I - A^k).$$

Letting $k \to \infty$ completes the proof of the assertion. #

COROLLARY. Given an $n \times n$ matrix A, for all complex z sufficiently close to 0, $(I - zA)^{-1}$ exists, and

$$(I - zA)^{-1} = \sum_{k=0}^{\infty} z^k A^k$$

in the sense of elementwise convergence.

Proof. Define δ by

$$\delta = \max_{i,j} |a_{ij}|.$$

Then putting $A^k = \{a_{ij}^{(k)}\}$, it follows from matrix multiplication that $|a_{ij}^{(2)}| \leqslant n\,\delta^2$, and, in general

$$|a_{ij}^{(k)}| \leqslant n^{k-1}\,\delta^k, \text{ all } i, j = 1, 2, \ldots, n.$$

Hence $A^k z^k \to 0$ if $|z| < (n\delta)^{-1}$, and the result follows from Lemma B.1. #

Note: We shall call the quantity $R(z) \equiv (I - zA)^{-1}$ the *resolvent* of A, for those z for which it exists, although this name is generally given to $(zI - A)^{-1}$.

The above Corollary also provides an analytical method for finding A^k for arbitrary k for real matrices of small dimension. It is necessary only to find the resolvent, and pick out the coefficient of z^k. Note that

$$(I - zA)^{-1} = \text{Adj}\,(I - zA)/\det\,(I - zA),$$

and the roots of $\det(I - zA) = z^n \det(z^{-1} I - A)$ for $z \neq 0$, are given by $z_i = \lambda_i^{-1}$ where the λ_i are the non-zero eigenvalues of A. Hence

$$\det(I - zA) = \prod_{j=1}^{n} (1 - z\,\lambda_j)$$

where the λ_j are the eigenvalues of A, and the singularities of the *resolvent* are all poles, possibly non-simple, since the elements of the *adjoint* matrix are polynomials of degree at most $n - 1$.

Example: Find P^k for the (stochastic) matrix:

$$P = \begin{bmatrix} 1 & 0 & 0 \\ p_1 & p_2 & p_3 \\ 0 & 0 & 1 \end{bmatrix}, \quad \begin{matrix} p_i > 0, i = 1, 2, 3 \\ \sum_{i=1}^{3} p_i = 1. \end{matrix}$$

$$(I - zP) = \begin{bmatrix} 1-z & 0 & 0 \\ -zp_1 & 1-zp_2 & -zp_3 \\ 0 & 0 & 1-z \end{bmatrix}$$

$$\det\,(I - zP) = (1 - z)^2 (1 - zp_2).$$

Therefore $(I - zP)^{-1}$

$$= (1-z)^{-2}(1-zp_2)^{-1} \begin{bmatrix} (1-z)(1-zp_2) & 0 & 0 \\ zp_1(1-z) & (1-z)^2 & zp_3(1-z) \\ 0 & 0 & (1-z)(1-zp_2) \end{bmatrix}.$$

By the use of partial fractions and the separating of components, we get eventually

$$(I - zP)^{-1} = (1-z)^{-1} \begin{bmatrix} 1 & 0 & 0 \\ p_1(1-p_2)^{-1} & 0 & p_3(1-p_2)^{-1} \\ 0 & 0 & 1 \end{bmatrix} +$$

$$+ (1-p_2 z)^{-1} \begin{bmatrix} 0 & 0 & 0 \\ -p_1(1-p_2)^{-1} & 1 & -p_3(1-p_2)^{-1} \\ 0 & 0 & 0 \end{bmatrix}.$$

Thus equating coefficients of z^k,

$$P^k = \begin{bmatrix} 1 & 0 & 0 \\ p_1(1-p_2)^{-1} & 0 & p_3(1-p_2)^{-1} \\ 0 & 0 & 1 \end{bmatrix} +$$

$$+ p_2^k \begin{bmatrix} 0 & 0 & 0 \\ -p_1(1-p_2)^{-1} & 1 & -p_3(1-p_2)^{-1} \\ 0 & 0 & 0 \end{bmatrix}.$$

LEMMA B.2. If $A = \{a_{ij}\}$ is a finite n x n matrix with real or complex elements, the matrix series

$$\sum_{k=0}^{\infty} t^k A^k / k!$$

converges elementwise; for each $t > 0$ the limit matrix is denoted by $\exp(tA)$, in analogy to the scalar case. If B is also an $(n \times n)$ matrix, then if $AB = BA$

$$\exp(A + B) = \exp(A) \cdot \exp(B).$$

Proof. Let $\delta = \max_{i,j} |a_{ij}|$. As in the Corollary to Lemma B.1

$$|a_{ij}^{(k)}| \leqslant n^{k-1} \delta^k, \text{ all } i, j = 1, 2, \ldots, n$$

and elementwise (absolute) convergence follows from the fact that

$$\sum_{k=1}^{\infty} t^k \frac{n^{k-1} \delta^k}{k!}$$

converges for any t, δ, $n > 0$.

The second part of the assertion follows from this simply by multiplying out the series for $\exp(A)$, $\exp(B)$, and collecting appropriate terms, as is permissible under the circumstances. #

Appendix C. Unitary transformations

DEFINITION C.1. An $n \times n$ matrix V with real or complex elements is said to be unitary if $\overline{V}' = V^{-1}$, where $\overline{V}$ is the complex conjugate of V (obtained by taking the complex conjugate of all its elements).

DEFINITION C.2. A unitary transformation of an $n \times n$ matrix A is a similarity transformation by a unitary matrix i.e. a transformation of the form $V^{-1} A V$, where V is unitary.

The reader should recall the well-known facts (1) that the eigenvalues (spectrum) of A remain invariant under similarity transformation; (2) that the diagonal elements of a triangular matrix are its eigenvalues.

LEMMA C.1. (Schur's Theorem). Any $n \times n$ matrix A of real or complex numbers can be transformed into an upper triangular matrix U by a suitable unitary transformation V. Moreover any specific eigenvalue of A (and hence of U) can be taken to be the (1,1) diagonal element of U.

Proof. This proceeds by induction on the dimension n of the matrix A.

The proposition is trivially true for $n = 1$; assume it is true for arbitrary $n \geqslant 1$. Now let A be an $(n + 1) \times (n + 1)$ matrix, λ any eigenvalue of it and $z = \{z_i\} \neq \mathbf{0}$ a corresponding right eigenvector, normed so that

$$z'\overline{z} = \sum_{i=1}^{n} |z_i|^2 = 1.$$

Now, it is possible to construct an $(n + 1) \times (n + 1)$ unitary matrix V^*_{n+1} with z as its first column (by considering an orthonormal basis including z as a member, of the vector space of all complex valued vectors with $n + 1$ elements). It then follows easily that

$$V^{*-1}_{n+1} A\, V^*_{n+1} = \begin{bmatrix} \lambda & s' \\ \mathbf{0} & B \end{bmatrix}$$

where B is some $(n \times n)$ matrix and s' a $(1 \times n)$ vector. By induction, there exists a unitary transformation matrix V_n such that

$$V_n^{-1} \, B \, V_n = U_n \text{ is upper triangular.}$$

Put
$$V_{n+1} = \begin{bmatrix} 1 & 0' \\ 0 & V_n \end{bmatrix}, \text{ so that } V_{n+1}^{-1} = \begin{bmatrix} 1 & 0' \\ 0 & V_n^{-1} \end{bmatrix}$$

Then
$$V_{n+1}^{-1} \, V_{n+1}^{*-1} \, A \, V_{n+1}^{*} \, V_{n+1} = \begin{bmatrix} \lambda & 0' \\ 0 & U_n \end{bmatrix}$$

which is upper triangular, with (1,1) element as required. It remains only to show that $V_{n+1}^* \, V_{n+1}$ is unitary; this follows trivially from the fact that both V_{n+1}^* and V_{n+1} are unitary. #

Unitary matrices have a special role in the vector space of all $(n \times 1)$ vectors of complex elements, which is sometimes known as unitary space, C_n.

The Euclidean *norm* or length of a member $z = \{z_i\}$ of this space is defined by

$$\|z\| = (\bar{z}'z)^{\frac{1}{2}} = (\sum_{i=1}^{n} |z_i|^2)^{\frac{1}{2}},$$

and the norm of a $(m \times n)$ matrix $A = \{a_{ij}\}$ regarded as a linear operator mapping vectors from C_n to C_m is defined by

$$\sup_{\|z\| \leqslant 1} \|A \, z\|.$$

It is well known and not difficult to see that for compatible matrices A and B,

$$\|A + B\| \leqslant \|A\| + \|B\|, \; \|AB\| \leqslant \|A\| \, \|B\|.$$

Moreover, if we put $K_A = \max_{i,j} |a_{ij}|$, it is useful to note the obvious inequalities:

$$K_A \leqslant \|A\| \leqslant (\sqrt{m}) n \, K_A$$

where A is $(m \times n)$ so that, e.g. elementwise convergence or uniform convergence to zero of a *set* of such matrices A is equivalent to such convergence in the Euclidean norm; as also e.g. boundedness, in the norm or in considering such matrices as points in the space of *mk*-coordinate vectors, are equivalent. (The left hand inequality is obtained by first putting z equal to the vector with a 1 in an arbitrary position, and zeroes elsewhere, the right hand inequality by use of the triangle inequality for complex numbers, in the definition of $\|A\|$.)

A unitary linear transformation: $z \to Vz$, where V is a unitary matrix, leaves the norm *invariant,* for

$$\|Vz\| = (\bar{z}'\,\bar{V}'Vz)^{\frac{1}{2}} = (\bar{z}'z)^{\frac{1}{2}} = \|z\|,$$

so that for any $n \times n$ A,

$$\|V^{-1}AV\| = \|V^{-1}A\| = \|A'\bar{V}\| = \|A\|$$

since $\|A'\| = \|A\|$, and $\bar{V}$ is unitary also, so that an unitary transformation of a square matrix leaves the Euclidean norm of it invariant.

Appendix D. Some real-variable theory

D.1. Upper semi-continuous functions

Let $f(x)$ be a mapping of a subset $\mathscr{A}$ of the Euclidean n-space R_n, into R_1 extended by the values $+\infty$ and $-\infty$.

DEFINITION D.1. The function f is said to be upper semicontinuous on $\mathscr{A}$, if

$$\limsup_{k\to\infty} f(x_k) \leqslant f(x_0)$$

for any $x_0 \in \mathscr{A}$, where $\{x_k\}$ is any sequence contained in $\mathscr{A}$ such that $x_k \to x_0$ as $k \to \infty$.

LEMMA D.1. (*a*) For any (finite, denumerable, or even non-countable) set of functions Λ which are upper semi-continuous on $\mathscr{A}$,

$$h(x) = \inf_{f\in\Lambda} f(x)$$

is also upper semi-continuous on $\mathscr{A}$.

(*b*) An upper semi-continuous function defined on a compact $\mathscr{A}$ attains its supremum for some $x_0 \in \mathscr{A}$.

Proof. (*a*) By definition for every $f(x) \in \Lambda$

$$h(x) \leqslant f(x), \ \forall x \in \mathscr{A}.$$

Thus for any $x_0 \in \mathscr{A}$ and $x_k \to x_0$, with $\{x_k\}$ in $\mathscr{A}$

$$h(x_k) \leqslant f(x_k), f \in \Lambda$$

so that

$$\limsup_{k\to\infty} h(x_k) \leqslant \limsup_{k\to\infty} f(x_k) \leqslant f(x_0)$$

for any $f \in \Lambda$.

Thus

$$\limsup_{k\to\infty} h(x_k) \leqslant \inf_{f\in\Lambda} f(x_0) = h(x_0)$$

(*b*) Let $a = \sup_{x \in \mathscr{A}} f(x)$ for an f upper-semicontinuous on $\mathscr{A}$. We can find a sequence $\{x_k\} \subset \mathscr{A}$ such that

$$f(x_k) \to a.$$

Since $\mathscr{A}$ is a compact, there exists a convergent subsequence $\{x_{k_i}\}$, $i = 1, 2, \ldots$ converging to a value $x_0 \in \mathscr{A}$. By upper semicontinuity of f

$$a = \limsup_{i \to \infty} f(x_{k_i}) \leqslant f(x_0) \leqslant a.$$

Thus

$$f(x_0) = a. \quad \#$$

D.2. Convexity, superconvexity, and Hölder's inequality

DEFINITION D.2. A function $f(x)$ is said to be convex on a fixed interval $\mathscr{I}$: $x_1 < x < x_2$ if for all $\alpha, \beta \geqslant 0$ such that $\alpha + \beta = 1$, and y, z in $\mathscr{I}$,

$$f(\alpha y + \beta z) \leqslant \alpha f(y) + \beta f(z).$$

DEFINITIONS D.3. A function $g(x)$ is said to be superconvex (log-convex) on a fixed interval $\mathscr{I}$: $x_1 < x < x_2$ if it is positive on this interval, and $\log g(x)$ is convex.

Thus, a superconvex function satisfies

$$g(\alpha y + \beta z) \leqslant g^{\alpha}(y)\, g^{\beta}(z)$$

where α, β, y and z are as in Definition D.2.

LEMMA D.2 (Hölder's Inequality for sums†). If $a_k, b_k \geqslant 0$, and $\alpha, \beta \geqslant 0$ and $\alpha + \beta = 1$, then

$$\sum_{k=1}^{n} a_k^{\alpha}\, b_k^{\beta} \leqslant \left(\sum_{k=1}^{n} a_k\right)^{\alpha} \left(\sum_{k=1}^{n} b_k\right)^{\beta}.$$

Proof. We note the elementary inequality

$$x^m - 1 \leqslant m(x-1) \text{ for } x \geqslant 1, 0 \leqslant m \leqslant 1,$$

which follows for $0 < m < 1$ and $x > 1$ from the Mean-Value Theorem applied to the function x^m on the interval $[1, x]$. Putting $x = a/b$ $(a \geqslant b)$, and multiplying by b

$$a^m\, b^{1-m} \leqslant b + m(a-b).$$

Putting $m = \alpha$, $\beta = 1 - m$,

$$a^{\alpha}\, b^{\beta} \leqslant \alpha a + \beta b \tag{D.1}$$

and since this is symmetric in a and b, it holds for any two positive numbers a, b. It clearly continues to hold if either a or b are permitted to become 0.

† Analogous inequalities for integrals may be found in almost any book dealing with the theory of functions.

The validity of the Lemma is trivial if all a_ks or all b_ks are zero, so we can exclude this case; then we have by (D.1)

$$\sum_{k=1}^{n}\left(\frac{a_k}{\sum_{k=1}^{n} a_k}\right)^{\alpha}\left(\frac{b_k}{\sum_{k=1}^{n} b_k}\right)^{\beta} \leqslant \sum_{k=1}^{n}\left(\alpha\frac{a_k}{\sum_{k=1}^{n} a_k}+\beta\frac{b_k}{\sum_{k=1}^{n} b_k}\right)$$

$$= \alpha + \beta = 1$$

so that the result of the Lemma follows by cross-multiplication. #

LEMMA D.3.† (i) The set of functions superconvex on $\mathscr{I}$ is closed under addition, multiplication and raising to any positive power.

(ii) If for each k, f_k is superconvex on $\mathscr{I}$, and $f = \lim\sup f_k$ is finite and positive on $\mathscr{I}$, then f is superconvex on I.

(iii) A function superconvex on $\mathscr{I}$ is convex on $\mathscr{I}$.

Proof. (i) Let f and g be superconvex on $\mathscr{I}$; then $h = f + g$ is positive on $\mathscr{I}$ and with α, β, y, z as specified before

$$h(\alpha y + \beta z) = f(\alpha y + \beta z) + g(\alpha y + \beta z)$$

$$\leqslant f^{\alpha}(y) f^{\beta}(z) + g^{\alpha}(y) g^{\beta}(z)$$

$$\leqslant (f(y) + g(y))^{\alpha} (f(z) + g(z))^{\beta}$$

by Hölder's Inequality (Lemma D.2)

$$= h^{\alpha}(y)\, h^{\beta}(z)$$

so that superconvexity of h follows by taking logarithms. Superconvexity under multiplication and raising to positive powers follows trivially from the fact that any positive linear combination of convex function is convex.

(ii) Let $f = \lim\sup_k f_k > 0$; then

$$f(\alpha y + \beta z) \leqslant \lim\sup f_k^{\alpha}(y) f_k^{\beta}(z)$$

$$\leqslant f^{\alpha}(y) f^{\beta}(z)$$

as required, the crucial step following from the superconvexity of f_k.

(iii) If f is superconvex on $\mathscr{I}$,

$$f(\alpha y + \beta z) \leqslant f^{\alpha}(y) f^{\beta}(z), \leqslant \alpha f(y) + \beta f(z)$$

the last inequality following from (D.1) in the proof of Lemma D.2. This yields convexity of f. #

† e.g. Kingman (1961).

Bibliography

The following list includes all titles referred to in the text, and a very few others.

Albert, E. (1963) Markov chains and λ-invariant measures. *J. Math. Anal. Applicns.*, **6**, 404–18.

Allen, B., Anderssen, R. S. & Seneta, E. (1973) Numerical investigation of infinite non-negative matrices. (Forthcoming Tech. Report, A.N.U.)

Bassett, L., Maybee, J. & Quirk, J. (1968) Qualitative economics and the scope of the correspondence principle. *Econometrica,* **36**, 544–63.

Bellman, R. (1960) *Introduction to Matrix Analysis.* McGraw-Hill, New York.

Bernstein, S. N. (1946) *Teoriya Veroiatnostei* (4th edition). Gostehizdat, Moscow–Leningrad. (See esp. pp. 203–13, 465–84, which are reprinted in Bernstein (1964).)

–– (1964) Classification of Markov chains and their matrices [in Russian]. *Sobranie Sochineniy: Tom IV, Teoriya Veroiatnostei i Matematicheskaia Statistika [1911–1946],* pp. 455–83. Izd, Nauka.

Birkhoff, G. (1946) Tres observaciones sobre el algebra lineal. *Univ. Nac. Tucumán Rev.*, Ser *A* **5**, 147–50.

Birkhoff, G. & Varga, R. S. (1958) Reactor criticality and non-negative matrices. *J. Soc. Industr. Appl. Math.*, **6**, 354–77.

Brauer, A. (1957*a*) The theorems of Ledermann and Ostrowski on positive matrices. *Duke Math. J.*, **24**, 265–74.

–– (1957*b*) A new proof of theorems of Perron and Frobenius on non-negative matrices, I. Positive matrices. *Duke Math. J.*, **24**, 367–78.

–– (1961) On the characteristic roots of power-positive matrices. *Duke Math. J.*, **28**, 439–45.

Brualdi, R. A., Parter, S. V. & Schneider, H. (1966) The diagonal equivalence of a non-negative matrix to a stochastic matrix. *J. Math. Anal. Applicns.*, **16**, 31–50.

Buchanan, M. L. & Parlett, B. N. (1966) The uniform convergence of matrix powers. *Numerische Math.*, **9**, 51–4.

Buharaev, R. G. (1968) Sets of processions of eigenvalues of stochastic matrices. *Soviet Math. Dokl.*, **9**, 64–7.

Burger, E. (1957) Eine Bemerkung über nicht-negative Matrizen. *Z. Angew. Math. Mech.*, **37**, 227.

Cherubino, S. (1957) *Calcolo Delle Matrici.* Edizione Cremonese, Roma.

Chung, K. L. (1967) *Markov Chains with Stationary Transition Probabilities* (2nd edition). Springer, Berlin.

Collatz, L. (1942) Einschliessungenssatz für die characteristischen Zahlen von Matrizen. *Math. Zeit.*, **48**, 221–6.

Daley, D. J. (1969) Quasi-stationary behaviour of a left-continuous random walk. *Ann. Math. Statist.*, **40**, 532–9.

Darroch, J. N. & Seneta, E. (1967) On quasi-stationary distributions in absorbing continuous-time finite Markov chains. *J. Appl. Prob.*, **4**, 192–6.

Debreu, G. & Herstein, I. N. (1953) Nonnegative square matrices. *Econometrica,* **21**, 597–607.

Derman, C. (1954) A solution to a set of fundamental equations in Markov chains. *Proc. Amer. Math. Soc.*, **5**, 332–4.

—— (1955) Some contributions to the theory of denumerable Markov chains. *Trans. Amer. Math. Soc.*, **79**, 541–55.

Dionísio, J. J. (1963/4, Fasc. 1) Matrizes não negativas. *Rev. Fac. Ciências Lisboa,* Ser A **10**, 5–35.

—— (1963/4, Fasc. 2) Sobre gráficos fortemente conexos e matrices não negativas. *Rev. Fac. Ciências Lisboa,* Ser. A **10**, 139–50.

Djoković, D. Ž. (1970) Note on non-negative matrices. *Proc. Amer. Math. Soc.*, **25**, 80–2.

Dmitriev, N. A. (1946) Characteristic roots of stochastic matrices [in Russian]. *Izvestiya A.N. S.S.S.R.* Ser. mat., **10**, 167–84.

Dmitriev, N. A. & Dynkin, E. B. (1945) On the characteristic numbers of stochastic matrices [in Russian]. *Doklady A.N. S.S.S.R.*, **49**, 159–62.

Doeblin, W. (1937) Le cas discontinu des probabilités en chaîne. *Publ. Fac. Sci. Univ. Masaryk* (Brno), no. **236**.

—— (1938) Exposé de la théorie des chaînes simples constantes de Markoff à un nombre fini d'états. *Revue Mathématique (Union Interbolkanique),* **2**, 77–105.

Doob, J. L. (1959) Discrete potential theory and boundaries. *J. Math. Mech.*, **8**, 433–58.

Dulmage, A. L. & Mendelsohn, N. S. (1962) The exponent of a primitive matrix. *Canad. Math. Bull.*, **5**, 241–4.

—— —— (1964) Gaps in the exponent set of primitive matrices. *Illinois J. Math.*, **8**, 642–6.

Erdös, P., Feller, W. and Pollard, H. (1949) A theorem on power series. *Bull. Amer. Math. Soc.*, **55**, 201–4.

Fan, K. (1958) Topological proofs for certain theorems on matrices with non-negative elements. *Monatsh. Math.* **62**, 219–37.

Feller, W. (1968) *An Introduction to Probability Theory and its Applications,* vol. 1 (3rd edition). Wiley, New York.

Fiedler, M. & Pták, V. (1962) On matrices with non-positive off-diagonal elements and positive principal minors. *Czechoslovak Math. J.*, **12**, 382-400.

Fréchet, M. (1938) *Recherches théoriques modernes sur le calcul des probabilités. Second livre. Méthode des fonctions arbitraires. Théorie des événements en chaîne dans le cas d'un nombre fini d'états possibles.* (Reprinted in 1952 with a new supplement and a note of Paul Lévy.) Gauthier-Villars, Paris.

Frobenius, G. (1908) Über Matrizen aus positiven Elementen. *S. – B. Preuss. Akad. Wiss.* (Berlin), 471–6; (1909), 514–18.

—— (1912) Über Matrizen aus nicht negativen Elementen. *S.–B. Preuss Akad. Wiss.* (Berlin), 456–77.

Gale, D. (1960) *The Theory of Linear Economic Models.* McGraw-Hill, New York.

Gantmacher, F. R. (1959) *Applications of the Theory of Matrices* (translated by J. L. Brenner). Interscience, New York.

Georgescu-Roegen, N. (1951) Some properties of a generalized Leontief model. Chapter 10 in *Activity Analysis of Production and Allocation*, ed. T. C. Koopmans. Wiley, New York.

Golub, G. & Seneta, E. (1973) Computation of the stationary distribution of an infinite Markov matrix. *Bull. Austral. Math. Soc.*, **8**, 333–41.

Gordon, P. (1965) *Théorie des Chaînes de Markov Finies et ses Applications.* Dunod. Paris.

Hadamard, J. (1928) Sur le battage des cartes et ses relations avec la Mécanique Statistique. *C.R. Congrés Int. des Mathématiciens de Bologne,* **5**, 133–9.

Hajnal, J. (1956) The ergodic properties of non-homogeneous finite Markov chains. *Proc. Cambridge Phil. Soc.*, **52**, 67–77.

—— (1958) Weak ergodicity in non-homogeneous Markov chains. *Proc. Cambridge Phil. Soc.*, **54**, 233–46.

Harris, T. E. (1957) Transient Markov chains with stationary measures. *Proc. Amer. Math. Soc.*, **8**, 937–42.

Hawkins, D. & Simon, H. A. (1949) Note: Some conditions of macroeconomic stability. *Econometrica,* **17**, 245–8.

Haynsworth, E. V. (1955) Quasi-stochastic matrices. *Duke Math. J.*, **22**, 15–24.

Heap, B. R. & Lynn. M. S. (1964) The index of primitivity of a non-negative matrix. *Numerische Math.*, **6**, 120–41.

—— —— (1966) The structure of powers of non-negative matrices. *J. SIAM Appl. Math.*, **14**, I: The index of convergence, 610–39; II: The index of maximum density, 762–77.

Heathcote, C. R. (1971) *Probability: Elements of the Mathematical Theory.* George Allen and Unwin, London.

Herstein, I. N. (1954) A note on primitive matrices. *Amer. Math. Monthly,* **61**, 18–20.

Hille, E. & Phillips, R. S. (1957) *Functional Analysis and Semi-Groups* (revised edition). American Math. Soc. Colloquium Publications, **31**, Providence, R. I.

Holladay, J. C. & Varga, R. S. (1958) On powers of non-negative matrices. *Proc. Amer. Math. Soc.*, **9**, 631–4.

Holmes, P. T. (1966) On some spectral properties of a transition matrix. *Sankhyā*, Ser. A **28**, 205–14; Corrigendum **29**, (1967), 106.

Hostinsky, B. (1931) *Méthodes générales du Calcul des Probabilités*, Mém. Sc. Math. no. 51. Gauthier-Villars, Paris.

Householder, A. S. (1958) The approximate solution of matrix problems. *J. Assoc. Comput. Mach.*, **5**, 204–43.

Howard, R. A. (1960) *Dynamic Programming and Markov Processes.* M.I.T. Press, Cambridge, Mass.

Joffe, A. & Spitzer, F. (1966) On multitype branching processes with $\rho \leqslant 1$. *J. Math. Anal. Applicns*, **19**, 409–30.

Karlin, S. (1959) *Mathematical Methods and Theory in Games, Programming, and Economics.* Addison-Wesley, Reading, Mass.

—— (1966) *A First Course in Stochastic Processes.* Academic Press, New York.

Karpelevich, F. I. (1949) On the characteristic roots of matrices with non-negative coefficients [in Russian] *Uspehi Mat. Nauk*, **4**, 177–8.

—— (1951) On the characteristic roots of matrices with non-negative elements [in Russian] *Izvestiya A.N. S.S.S.R.* Ser. mat., **15**, 361–83.

Kaucky, J. (1930) Quelques remarques sur les chaînes de Markoff [in Czech, French summary]. *Publ. Fac. Sci. Univ. Masaryk* (Brno), no. **131**.

Kemeny, J. G. (1966) Representation theory for denumerable Markov chains. *Trans. Amer. Math. Soc.*, **125**, 47–62.

Kemeny, J. G. & Snell, L. J. (1960) *Finite Markov Chains.* Van Nostrand, Princeton, N.J.

Kemeny, J. G., Snell, L. J. & Knapp, A. W. (1966) *Denumerable Markov Chains.* Van Nostrand, Princeton, N.J.

Kendall, D. G. (1959) Unitary dilations of Markov transition operators, and the corresponding integral representations for transition-probability matrices. In *Probability and Statistics* (Cramér Memorial Volume), ed. U. Grenander. Almqvist and Wiksell, Stockholm and New York, pp, 139–61.

—— (1960) Geometric ergodicity and the theory of queues. In *Mathematical Methods in the Social Sciences*, ed. K. J. Arrow, S. Karlin, P. Suppes. Stanford.

—— (1966) Contribution to discussion in 'Quasi-stationary distributions and time-reversion in genetics' by E. Seneta [with discussion] *J. Roy. Statist. Soc.*, Ser. B **28**, 253–77.

Khintchine, A. Y. (1969) *Mathematical Methods in the Theory of Queueing* (2nd edition) (translation D. M. Andrews and M. H. Quenouille, with additional notes by E. Wolman). Griffin, London.

Kingman, J. F. C. (1961) A convexity property of positive matrices. *Quart. J. Math., Oxford (2)*, **12**, 283–4.

—— (1963) The exponential decay of Markov transition probabilities. *Proc. London Math. Soc.*, **13**, 337–58.

Kolmogorov, A. N. (1931) Über die analytischen Methoden in der Wahrscheinlichkeitsrechnung, *Math. Ann.*, **104**, 415–58.

—— (1936) Anfangsgründe der Theorie det Markoffschen Ketten mit unendlichen vielen möglichen Zuständen. *Mat. Sbornik* (N.S.), **1**, 607–10; *Bull. Univ. Moscou* [in Russian], **1** (1937), no. 3, 1–16.

Konečný, M (1931) Sur la théorie des chaînes de Markoff. *Publ. Fac. Sci. Univ. Masaryk* (Brno), no. **147**.

Kotelyanskii, D. M. (1952) Some properties of matrices with positive elements [in Russian]. *Mat. Sbornik,* **31**, 497–506.

Kozniewska, I. (1962) Ergodicité et stationnarité des chaînes de Markoff variables à un nombre fini d'états possibles. *Colloq. Math.*, **9**, 333–46.

Kuich, W. (1970*a*) On the entropy of context-free languages. *Information and Control,* **16**, 173–200.

—— (1970*b*) On the convergence parameter of infinite non-negative matrices. *Monatsh. Math.*, **74**, 138–44.

Lappo-Danilevskii, J. A. (1934) Mémoires sur la théorie des systèmes des équations différentielles linéaires. *Trav. Inst. Phys.–Math. Stekloff. Acad. Sci. U.R.S.S.*, **1**. [All 3 volumes of Lappo-Danilevskii's work on the same topic, of which this is the first, were reprinted by Chelsea, N.Y.]

Larisse, J. (1966) Sur une convergence presque sûre d'un produit infini de matrices stochastiques ayant le même nombre de classes ergodiques. *C.R. Acad. Sci. Paris,* **262**, 913–15.

Larisse, J. & Schützenberger, M. P. (1964) Sur certaînes chaînes de Markov non homogènes. *Publ. Inst. Statist. Univ. Paris,* **8**, 57–66.

Lendermann, W. (1950*a*) On the asymptotic probability distribution for certain Markoff processes. *Proc. Cambridge Phil. Soc.*, **46**, 581–94.

—— (1950*b*) Bounds for the greatest latent root of a positive matrix. *J. London Math. Soc.*, **25**, 265–8.

Lopez, A. (1961) *Problems in Stable Population Theory.* Office of Population Research, Princeton University, Princeton.

McFarland, D. D. (1969) On the theory of stable populations: a new and elementary proof of the theorems under weaker assumptions. *Demography,* **6**, 301–22.

Malécot, G. (1944) Sur un problème de probabilités en chaîne que pose la génétique. *C.R. Acad. Sci. Paris,* **219**, 379–81.

Mandl, P. (1960) On the asymptotic behaviour of probabilities within groups of states of a homogeneous Markov process [in Russian]. *Čas. Pěst. Mat.*, **85**, 448–56.

—— (1967) An iterative method for maximizing the characteristic root of positive matrices. *Rev. Roum. Math. Pures et Appl.*, **12**, 1317–22.

Mandl, P. & Seneta, E. (1969) The theory of non-negative matrices in a dynamic programming problem. *Austral. J. Statist.*, **11**, 85–96.

Mangasarian, O. L. (1971) Perron–Frobenius properties of $Ax - \lambda Bx$. *J. Math. Anal. Applicns,* **36**, 86–102.

Marcus, M. & Minc, H. (1964) *A Survey of Matrix Theory and Matrix Inequalities.* Allyn and Bacon, Rockleigh, N.J.

—— —— (1967) On a conjecture of B. L. van der Waerden. *Proc. Cambridge Phil. Soc.,* **63**, 305–9.

Markov, A. A. (1907) Investigation of an important case of dependent trials [in Russian]. *Izvestiya Akad. Nauk S.P.–B. (6),* **1**, 61–80.

—— (1924) *Ischislenie Veroiatnostei* (4th edition) (esp. pp. 552–81). Gosizdat, Moscow.

Maybee, J. S. (1967) Matrices of class $\mathcal{J}_2$. *J. Res. Nat. Bur. Standards,* **71**B, 215–24.

Maybee, J. S. & Quirk, J. (1969) Qualitative problems in matrix theory. *SIAM Reveiw,* **11**, 30–51.

Metzler, L. A. (1945) A multiple-county theory of income transfers. *J. Political Economy,* **59**, 14–29.

—— (1951) Taxes and subsidies in Leontief's input–output model. *Quart. J. Economics,* **65**, 433–8.

Miller, H. D. (1961) A convexity property in the theory of random variables defined on a finite Markov chain. *Ann. Math. Statist.,* **32**, 1260–70.

Minkowski, H. (1900) Zur Theorie der Einheiten in den algebraischen Zahlkorpern. *Gesammelte Abhandlungen,* vol. 1 (1911), 316–19.

Mirsky, L. (1963) Results and problems in the theory of doubly-stochastic matrices. *Z. Wahrscheinlichkeitstheorie,* **1**, 319–34.

—— (1964) Inequalities and existence theorems in the theory of matrices. *J. Math. Anal. Applicns.,* **9**, 99–118.

von Mises, R. (1931) *Wahrscheinlichkeitsrechnung.* Fr. Deuticke, Leipzig and Vienna.

Morgenstern, O. (ed.) (1954) *Economic Activity Analysis.* Wiley, New York.

Morishima, M. (1952) On the laws of change of the price system in an economy which contains complementary commodities. *Osaka Economic Papers,* **1**, 101–13.

—— (1961) Generalizations of the Frobenius–Wielandt theorems for non-negative square matrices. *J. London Math. Soc.,* **36**, 211–20.

—— (1964) *Equilibrium, Stability, and Growth,* Clarendon Press, Oxford.

Mott, J. L. (1957) Conditions for the ergodicity of non-homogeneous finite Markov chains. *Proc. Roy. Soc. Edinburgh,* section A **64**, 369–80.

Mott, J. L. & Schneider, H. (1957) Matrix norms applied to weakly ergodic Markov chains. *Archiv der Mathematik,* **8**, 331–3.

Moy, S.–T. C. (1965) λ-continuous Markov chains. *Trans. Amer. Math. Soc.,* **117**, 68–91; Part II *ibid.* **120**, 83–107.

—— (1967*a*) Ergodic properties of expectation matrices of a branching process with countably many types. *J. Math. Mech.,* **16**, 1207–26.

—— (1967*b*) Extension of a limit theorem of Everett, Ulam and Harris on multitype branching processes to a branching process with countably many types. *Ann. Math. Statist.,* **38**, 992–9.

Nelson, E. (1958) The adjoint Markoff process. *Duke Math. J.,* **25**, 671–90.

Neveu, J. (1964) Chaînes de Markov et théorie du potentiel. *Ann. Fac. Sci. Univ. Clermont–Ferrand.,* **24**, 3(Math.), 37–89.

Oldenburger, R. (1940) Infinite powers of matrices and characteristic roots. *Duke Math. J.*, **6**, 357–61.

de Oliveira, G. N. (1968) *Sobre Matrizes Estocasticas e Duplamente Estócasticas.* Coimbra. (Also as: *Rev. Fac. Ci. Univ. Coimbra,* **41**, 15–221)

Orey, S. (1962) An ergodic theorem for Markov chains. *Z. Wahrscheinlichkeitstheorie,* **1**, 174–6.

Ostenc, É. (1934) Sur le principe ergodique dans les chaînes de Markov à elements variables. *C.R. Acad. Sci. Paris,* **199**, 175–6.

Ostrowski, A. M. (1937) Über die Determinanten mit überwiegender Hauptdiagonale. *Comment. Math. Helv.*, **10**, 69–96.

—— (1952) Bounds for the greatest latent root of a positive matrix. *J. London Math. Soc.*, **27**, 253–6.

—— (1956) Determinanten mit überwiegender Hauptdiagonale und die absolut Konvergenz von linearen Iterationsprozessen. *Comment. Math. Helv.*, **30**, 175–210.

Ostrowski, A. M. & Schneider, H. (1960) Bounds for the maximal characteristic root of a non-negative irreducible matrix. *Duke Math. J.*, **27**, 547–53.

Parthasarathy, K. R. (1967) *Probability Measures on Metric Spaces.* Academic Press, New York.

Paz, A. (1963) Graph-theoretic and algebraic characterizations of some Markov processes. *Israel J. Math.*, **1**, 169–80.

—— (1965) Definite and quasidefinite sets of stochastic matrices. *Proc. Amer. Math. Soc.*, **16**, 634–41.

Perfect, H. (1952) On positive stochastic matrices with real characteristic roots. *Proc. Cambridge Phil. Soc.*, **48**, 271–6.

—— (1953) Methods of constructing certain stochastic matrices. *Duke Math. J.*, **20**, 395–404; Part II, *ibid* **22** (1955), 305–11.

Perfect, H. & Mirsky, L. (1965) Spectral properties of doubly–stochastic matrices. *Monatsh. Math.* **69**, 35–57.

Perkins, P. (1961) A theorem on regular matrices. *Pacific J. Math.*, **11**, 1529–33.

Perron, O. (1907) Zur Theorie der Matrizen. *Math. Ann.*, **64**, 248–63.

Pruitt, W. E. (1964) Eigenvalues of non-negative matrices. *Ann. Math. Statist.*, **35**, 1797–800.

Pták, V. (1958) On a combinatorial theorem and its application to non-negative matrices [in Russian and English]. *Czechoslovak Math. J.*, **8**, 487–95.

Pták, V. & Sedláček, J. (1958) On the index of imprimitivity of non-negative matrices [in Russian and English]. *Czechoslovak Math. J.*, **8**, 496–501.

Putnam, C. R. (1958) On bounded matrices with non-negative elements. *Canad. J. Math.*, **10**, 587–91.

—— (1961) A note on non-negative matrices. *Canad. J. Math.*, **13**, 59–62. (A correction to the 1958 paper.)

Quine, M. P. (1972) The multitype Galton–Watson process with ρ near 1. *Adv. Appl. Prob.*, **4**, 429–52.

Romanovsky, V. I. (1936) Recherches sur les chaînes de Markoff *Acta Math.*, **66**, 147–251.

—— (1949) *Diskretnii Tsepi Markova.* G.I.T.–T.L. Moscow–Leningrad. (Translated in 1970 as *Discrete Markov Chains,* by E. Seneta. Wolters-Noordhoff, Groningen.)

Rosenblatt, D. (1957) On the graphs and asymptotic forms of finite Boolean relation matrices and stochastic matrices. *Naval Res. Logist. Quart.*, **4**, 151–67.

Samelson, H. (1957) On the Perron–Frobenius theorem. *Michigan Math. J.*, **4**, 57–9.

Sarymsakov, T. A. (1945) Sur une synthèse des deux méthodes d'exposer la théorie des chaînes discrètes de Markoff. *C.R. (Doklady) Acad. Sci. U.R.S.S.*, **48**, 159–61.

—— (1953*a*) On the ergodic principle for inhomogeneous Markov chains [in Russian]. *Doklady A.N. S.S.S.R.*, **90**, 25–8.

—— (1953*b*) On the matrix method of investigation of Markov chains with a countable set of possible states [in Russian]. *Sredneaziatskii Gos. Univ.: Trudy,* Novaia Seria, vyp. **46**, Fiz.–mat. nauki, kniga 9, pp. 1–46. Izd. S.–A. G.U. Tashkent.

—— (1954) *Osnovi Teorii Processov Markova.* G.I. T.–T.L., Moscow.

—— (1956) On the theory of inhomogeneous Markov chains [in Russian]. *Doklady A. N. Uzbek. S.S.R.*, **8**, 3–7.

—— (1958) On inhomogeneous Markov chains. *Doklady A.N. S.S.S.R.*, **120**, 465–7.

—— (1961) Inhomogeneous Markov chains [in Russian]. *Teor. Veroiat. Primenen.*, **6**, 194–201.

Sarymsakov, T. A. & Mustafin, H. A. (1957) On an ergodic theorem for inhomogeneous Markov chains [in Russian]. *Sredneaziatskii Gos. Univ. im V.I. Lenina: Trudy,* Novaia seria, vyp. **74**, Fiz-mat. nauki, kniga 15, pp. 1–38. Izd. A. N. Uzbekskoi S.S.R., Tashkent.

Schneider, H. (1958) Note on the fundamental theorem of irreducible non-negative matrices. *Proc. Edinburgh Math. Soc.*, **2**, 127–30.

Schwartz, Š. (1967) Some estimates in the theory of non-negative matrices. *Czechoslovak Math. J.*, **17**, 399–407.

Seneta, E. (1967*a*) Finite approximations to infinite non-negative matrices. *Proc. Cambridge Phil. Soc.*, **63**, 983–992; Part II: refinements and applications. *ibid.* **64**, (1968*a*) 465–470.

—— (1967*b*) On the maxima of absorbing Markov chains. *Austral. J. Statist.*, **9**, 93–102.

—— (1968*b*) The principle of truncations in applied probability. *Comm. Math. Univ. Carolinae,* **9**, 533–9.

—— (1973*a*) On strong ergodicity of inhomogenous products of finite stochastic matrices. *Studia Math.*, **46**, No. 3 (forthcoming).

—— (1973*b*) On the historical development of the theory of finite inhomogeneous Markov chains. *Proc. Cambridge Phil. Soc.* (to appear).

Seneta, E. & Vere-Jones, D. (1966) On quasi-stationary distributions in discrete–time Markov chains with a denumerable infinity of states. *J. Appl. Prob.*, **3**, 403–34.

Šidák, Z. (1962) Eigenvalues of operators in denumerable Markov chains. *Trans. 3rd Prague Conf. on Inf. Theory, Stat. Decision Fns., Random Processes.* (Published in 1964 by Publishing House of the Czechoslovak Acad. Sciences, Prague, pp. 641–56.)

–– (1963) Někvteré věty a příklady z teorie operátorů ve spočetných Markovových řetezcích. *Čas. Pěst. Mat.*, **88**, 457–78.

–– (1964*a*) Eigenvalues of operators in ℓ_p-spaces in denumerable Markov chains. *Czechoslovak Math. J.*, **14**, 438–43.

–– (1964*b*) O počtu kladných prvků v mocninách nezáporné matice. *Čas Pěst Mat.*, **89**, 28–30.

Sinkhorn, R. (1964) A relationship between arbitrary positive matrices and doubly stochastic matrices. *Ann. Math. Statist.*, **35**, 876–9.

Sinkhorn, R. & Knopp, P (1967) Concerning non-negative matrices and doubly stochastic matrices. *Pacific J. Math.*, **21**, 343–8.

Sirazhdinov, S. H. (1950) The ergodic principle for non-homogeneous Markov chains [in Russian]. *Doklady A.N. S.S.S.R.*, **71**, 829–30.

Solow, R. (1952) On the structure of linear models. *Econometrica*, **20**, 29–46.

Solow, R. & Samuelson, P. A. (1953) Balanced growth under constant returns to scale. *Econometrica*, **21**, 412–24.

Stein, P. & Rosenberg, R. L. (1948) On the solution of linear simultaneous equations by iteration. *J. London Math. Soc.*, 23, 111–18.

Suleĭmanova, H. R. (1949) Stochastic matrices with real characteristic numbers [in Russian]. *Doklady A.N. S.S.S.R.*, **66**, 343–5.

–– (1953) On the characteristic numbers of stochastic matrices [in Russian]. *Uchenie Zap. Mosk. Ped. Inst.*, **71**, 167–77.

Tambs-Lyche, R. (1928) Un théorème sur les déterminants. *Det. Kong. Vid. Selskab.*, Forh. I, nr. 41, 119–20.

Taussky, O. (1949) A recurring theorem on determinants. *Amer. Math. Monthly.*, **56**, 672–6.

Thomasian, A. J. (1963) A finite criterion for indecomposable channels. *Ann. Math. Statist.*, **34**, 337–8.

Timan, O. Z. (1972) On the boundaries of Brauer for characteristic numbers of indecomposable stochastic matrices [in Russian]. *Zh. Vychisl. Mat. i Mat. Fiz.*, **12**, 192–8.

Titchmarsh, E. C. (1939) *The Theory of Functions* (2nd edition). Oxford University Press, London.

Tweedie, R. L. (1971) Truncation procedures for non-negative matrices. *J. Appl. Prob.*, **8**, 311–20.

Ullman, J. L. (1952) On a theorem of Frobenius. *Michigan Math. J.*, **1**, 189–93.

Varga, R. S. (1962) *Matrix Iterative Analysis.* Prentice-Hall, Englewood Cliffs, N.J.

Veech, W. (1963) The necessity of Harris' condition for the existence of a stationary measure. *Proc. Amer. Math. Soc.*, **14**, 856–60.

Vere-Jones, D. (1962) Geometric ergodicity in denumerable Markov chains. *Quart. J. Math. Oxford (2)*, **13**, 7–28.

—— (1963) On the spectra of some linear operators associated with queueing systems. *Z. Wahrscheinlichkeitstheorie*, **2**, 12–21.

—— (1966) Note on a theorem of Kingman and a theorem Chung. *Ann. Math. Statist.*, **37**, 1844–6.

—— (1967) Ergodic properties of non-negative matrices – I. *Pacific J. Math.*, **22**, 361–86; – II *ibid.* **26** (1968), 601–20.

—— (1971) Finite bivariate distributions and semigroups of non-negative matrices. *Quart. J. Math. Oxford (2)*, **22**, 247–70.

van der Waerden, B. L. (1926) Aufgabe 45. *Jber. Deutsch. Math. Verein.*, **35**, 117.

Wielandt, H. (1950) Unzerlegbare, nicht negative Matrizen. *Math. Zeit.*, **52**, 642–8.

Wolfowitz. J. (1963) Products of indecomposable, aperiodic, stochastic matrices. *Proc. Amer. Math. Soc.*, **14**, 733–7.

Wong, Y. K. (1954) Inequalities for Minkowski–Leontief matrices. In: *Economic Activity Analysis* (ed. O. Morgenstern), pp. 201–81. Wiley, New York.

Supplementary note on Bibliography

The following relevant references have come to the author's attention more recently:

Keilson, J. H. and Styan, G. P. H. (1973) Markov chains and M-matrices: inequalities and equalities. *J. Math. Anal. Applicns.*, **41**, 439–59. (to Bibliography and discussion to §2.3.)

Klimko, L. A. and Sucheston, L. (1968) On probabilistic limit theorems for a class of positive matrices. *J. Math. Anal. Applicns.*, **24**, 191–201. (to Bibliography and discussion to Chapters 5, 6.)

Medlin, G. W. (1953) On bounds for the greatest characteristic root of a matrix with positive elements. *Proc. Amer. Math. Soc.*, **4**, 769–71. (to §2.5.)

de Oliveira, G. N. (1972) On the characteristic vectors of a matrix. *Linear Algebra and Applicns.*, **5**, 189–96. (to §2.5.)

Paz, A. (1971) *Introduction to Probabilistic Automata.* Academic Press, New York. (to §4.3, and its Bibliography and discussion.)

Rheinboldt, W. C. and Vandergraft, J. S. (1973) A simple approach to the Perron-Frobenius theory for positive operators on general partially-ordered finite-dimensional linear spaces. *Mathematics of Computation*, **27**, 139–45. (to Bibliography and discussion to §2.3.)

Robert, F. (1973) *Matrices Non-negatives et Normes Vectorielles* (cours de DEA) (mimeographed lecture notes). (To Bibliography and discussion of §§2.2, 2.3.)

Glossary of notation and symbols

T	usual notation for a non-negative matrix.
A	typical notation for a matrix.
A'	the transpose of the matrix A.
a_{ij}	the (i, j) entry of the matrix A.
$\tilde{T}$	the incidence matrix of the non-negative matrix T.
P	usual notation for a stochastic matrix.
0	zero; the zero matrix.
$\underset{\sim}{x}$	typical notation for a column vector.
$\underset{\sim}{0}$	the column vector with all entries 0.
$\underset{\sim}{1}$	the column vector with all entries 1.
$P_1 \sim P_2$	the matrix P_1 has the same incidence matrix as the matrix P_2.
$\min^+$	the minimum among all strictly positive elements.
R_k	k-dimensional Euclidean space; a certain submatrix.
R	set of strictly positive integers; convergence parameter of an irreducible matrix T; a certain submatrix.
I	the identity (unit) matrix; the set of inessential indices.
$i \in R$	i is an element of the set R.
$\mathscr{C} \subset \mathscr{S}$ or $\mathscr{C} \subseteq \mathscr{S}$	$\mathscr{C}$ is a subset of the set $\mathscr{S}$.
$_{(n)}T$	$(n \times n)$ northwest corner truncation of T.
$\Delta_i, \Delta_i(s)$	the principal minor of $(sI - T)$.
$_{(n)}\Delta(\beta)$	$\det [_{(n)}I - \beta \cdot {}_{(n)}T]$.
$_{(n)}\Delta$	$_{(n)}\Delta(R)$
M.C.	Markov chain.
$\mathscr{E}$	mathematical expectation operator.
G_1	class of $(n \times n)$ regular matrices.
M	class of $(n \times n)$ Markov matrices.
G_2	class of stochastic matrices defined on p. 106.
G_3	class of $(n \times n)$ scrambling matrices.

Author index

Subject index